SAWDUST EMPIRE

The forest: massive old-growth pines, towering above the forest floor. *Courtesy of Stephen F. Austin State University, Forest History Collections.*

Sawdust Empire

THE TEXAS LUMBER INDUSTRY,

1830–1940

by

ROBERT S. MAXWELL

and

ROBERT D. BAKER

TEXAS A&M UNIVERSITY PRESS

COLLEGE STATION

Library of Congress Cataloging in Publication Data

Maxwell, Robert S.
    Sawdust empire.

    Bibliography: p.
    Includes index.
        1. Lumber trade—Texas—History. 2. Lumbering—
Texas—History. 3. Logging—Texas—History. 4. Forest
conservation—Texas—History.   I. Baker, Robert D. II. Title.
HD9757.T45M38   1983   338.4′7674′09764   82-40442
ISBN 0-89096-148-4 (cloth)
ISBN 1-58544-059-0 (pbk.)

Manufactured in the United States of America
First Paperback Edition

For Elizabeth

Contents

List of Illustrations

Preface

THERE has long been a need for a history of the Texas lumber industry. Before 1901 lumbering was the state's largest manufacturing enterprise, first among Texas industries in generating income and the largest single employer. Until 1930 lumbering continued to hold second place (after oil) in these categories among manufacturing industries of the state. For thousands of Texans lumbering was a way of life, and for a few it was the route to great riches and vast power. Yet except for some specialized accounts, biographical sketches, and short articles, the history of the Texas lumber industry has been largely ignored.

The story begins before 1830 with the earliest white settlers, who laboriously transformed lumber, one board at a time, into frontier dwellings. The commercial industry began about 1880, and the following half-century was truly a bonanza era, marked by the rise, peak, and decline of Texas lumbering based on the exploitation of the virgin pine forests west of the Sabine. The struggle to cope with the Great Depression, the entrance of the federal government into Texas forestry, and the first technological breakthrough in diversified pine usage brings the story to 1940, the eve of World War II, and the end of this study.

The present volume is based on more than twenty years of research in the company records of some of the largest and oldest lumbering firms in the state. Several dozen veteran lumbermen, ranging from owners to loggers and professional foresters, have been interviewed and their comments recorded. The available secondary books and periodical literature have also been consulted. This study

has tried to recreate the story of the world of Texas lumbering as it existed before World War II. We have sought to understand and interpret the industry and its people, both owners and workers, in the light of their own standards and mores.

The history of the Texas lumber industry is a fascinating and complex story involving hundreds of operators at any given time. In the course of the years covered by this study thousands of individuals and companies engaged in logging and sawmilling. Some were small; others, larger and more stable, spanned much of the entire era. Many lumbermen left their imprint in the form of place names—towns, streets, buildings, and parks—in East Texas. Several hundred young men and women bear names that indicate their parents' admiration for the lumber tycoons of the bonanza period. In short, the lumber industry and its personalities dominated both the region and the era.

In discussing the development of the Texas lumber industry over the course of more than a century, it would be neither possible nor desirable to mention every person or company in the business. Those selected as representative of the industry have mostly been those organizations that have remained in operation more than one generation and have contributed to the stability of the region. If large-scale operators have been discussed more than the small mills, it is because the latter were more ephemeral and left fewer records. Then, too, it was the large operator who spoke for the industry, belonged to the trade associations, and set the pattern of policies for the entire industry. But both the small sawmillers and the hired workers have also been studied in some detail. The account of the beginnings of the struggle for conservation of forest resources and of the work of the early conservation leaders presents an important aspect of the history of Texas forests. Without their efforts the vigorous second-growth forest that flourished in the generation after World War II might never have existed.

The end of the study with the developments leading up to 1940 marks a natural division point. The half-century or so before that date witnessed the development, the peak years of maximum production, the decline, and the struggle for survival of the traditional lumber industry based on perfecting the big mill. That year also marked the passing of the last of the first generation of lumber giants, whose names were synonymous with the industry. The war years and the generation after the war brought about a new forest products technology, including advances in wood chemistry, lamination, automation, diversification, and new inventions that fostered a much more complex and technical industry. The study of these changes and the new forest products industry should furnish the stuff for a new and different book planned for the future.

Acknowledgments

MANY people have contributed to the completion of this book. At Stephen F. Austin State University three presidents have, in turn, given encouragement and support; the university has provided a series of grants for travel and study; and the successive directors of the SFA Library have been most cooperative in collecting and organizing the materials of the Forest History Collections. The members of the SFA School of Forestry, especially Laurence C. Walker and Kent Adair, provided valued professional assistance in many aspects of the project.

Thanks are due to directors Elwood Maunder and Harold Steen and to editor Ronald Fahl of the Forest History Society, who have helped in many areas and have contributed a number of pictures not available elsewhere. Ralph Hidy and Arthur M. Johnson, then of the Harvard School of Business Administration, helped to delimit the subject and to block out the outlines for this study. Loyd Collier and Bruce Lyndon Cunningham have contributed original maps and sketches that are much appreciated. My daughter, Elizabeth Maxwell; Douglas Davis, of the Forest History Society; and George K. Stephenson, United States Forest Service, retired, have read all or parts of the manuscript and offered valuable suggestions.

My colleague, Robert D. Baker, has been active in providing ideas and professional forestry knowledge, and his associate, Robert C. Maggio, has helped manage the pictures, maps, and charts for the work. Finally, special thanks are due to Doris Goodrich Jones and to Dr. Luther G. Jones for their generous assistance in underwriting the cost of the illustrations.

Robert S. Maxwell
November 1, 1982

SAWDUST EMPIRE

Texas forests as they were. The old-growth stand
shows a clear, parklike forest floor. *Authors'*
collections.

CHAPTER I

The Forest and the People

The countrey is overgrowne with pyne.—
Captain John Smith

THE Texas pine forest was impressive—a magnificent, towering, massive virgin stand. It contained several species of the genus *Pinus*: the lordly longleaf, the shortleaf, the loblolly, and perhaps an occasional hybrid combination. These species were intermixed, especially in the northern and western reaches of the forest, with many hardwoods—oaks, gums, ash, and others. Cypress grew in and along the rivers, especially the lower Sabine, but throughout the region the pines were dominant and were the principal commercial timber. Indeed, the Texas pine forest was but the westernmost extension of the great trans-Mississippi evergreen forest, embracing Arkansas, eastern Oklahoma, and Louisiana, as well as East Texas. It reached from southern Missouri to the flatlands of the Gulf of Mexico and from the Mississippi to the broad blackland prairies of Central Texas.

One writer described the westernmost extension and termination of the southern pine forest in dramatic terms, likening it to an ocean wave on a beach:

It is a striking phenomenon, this breaking up and gradual dwindling away of so vast and vigorous a forest. . . . Like a vast wave that has rolled in upon a level beach, the Atlantic forest breaks upon the dry plains—halting, creeping forward, thinning out, and finally disappearing, except where, along a river course, it pushes far inland.[1]

It was one great forest until man arrived and subdivided it with artificial lines called state boundaries. These political divisions have given the pines of Texas a different history, though not a dissimilar one, from that of their neighbors. It is essentially the Texas story that this book investigates, but some mention will be made of the related activities and operations east of the Sabine.

The forest region of East Texas includes all or parts of forty-eight counties and encompasses more than 36,000 square miles. Thus the area is as large as the state of Indiana and larger than any of twelve other states. It extends from the Gulf coastal plains on the south to the Red River and from the Sabine to an irregular line somewhat short of the 96th meridian, where the pine forests thin out and give way to scattered groves and open prairie.[2] The commercial-forest region is considerably smaller, comprising 14 to 18 million acres, the concentration of trees per acre becoming denser as one proceeds south and east. Described in another way, the East Texas pine forest is densest (having the largest individual trees and the most trees per acre) near the Louisiana boundary, where the annual rainfall approaches fifty inches. The stands become lighter and more open as one goes north and west, and the forest disappears entirely at the line

[1] William L. Bray, *Forest Resources of Texas*, U.S. Department of Agriculture, Bureau of Forestry Bulletin no. 47 (1904), p. 15.

[2] Charles Mohr, *The Timber Pines of the Southern United States*, U.S. Department of Agriculture, Division of Forestry Bulletin no. 18 (1897); S. B. Buckley, "Pine Lands of South-eastern Texas," *Texas Almanac*, 1868, p. 92.

where the annual precipitation dips below forty inches. It is a large area, comprising more than 10 percent of Texas. In the early days, however, it was little known, a sharp contrast to the popular stereotype of Texas plains, and was considered somewhat remote by residents of the urban centers of the state.[3] Although there is a considerable variety in the merchantable hardwoods of East Texas, the dominant species is the pine. And pine lumbering constitutes about 85 percent of the logging activity of the region.

The longleaf area, originally including about five thousand square miles, thrust itself like a broad wedge southwestward from the Louisiana border to the Trinity River between the loblolly on the south and the shortleaf on the north.[4] By no means a solid stand, the longleaf forests included parts of present-day Sabine, San Augustine, Nacogdoches, Angelina, Polk, Trinity, Tyler, Jasper, Newton, Orange, and Hardin counties. The typical longleaf country was relatively open. The sandy soil was well drained and of a texture that permitted the pine to seek the sun with its crown while thrusting its taproot deep into the water-bearing sands so as to be relatively independent of surface moisture conditions. On such terrain the longleaf flourished, and, having achieved dominance, had choked out the underbrush and scrub hardwoods, with the assistance of wood fires, thus producing the parklike forest floor described by the early travelers in the region. In depressions and poorly drained areas hardwoods penetrated into the longleaf forest. The longleaf pine reproduced with difficulty in the face of even its natural competition, and the logging methods of the early period left but a scant basis for perpetuation of the species. Consequently, most of the longleaf stands have disappeared and have been replaced by loblolly forests or stands of mixed loblolly and hardwood.[5]

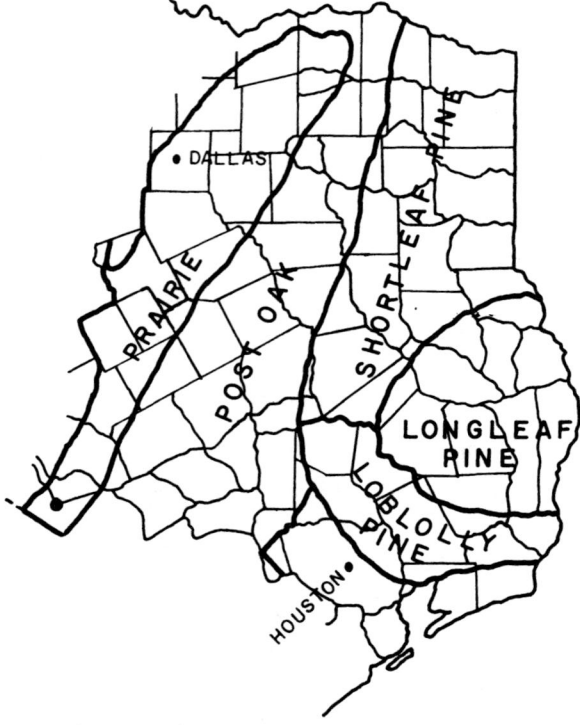

Distribution of timber varieties in East Texas. *Authors' collections.*

As a general pattern the loblolly forest lay south and west of the longleaf stands. It extended northward from the Gulf coastal plain and intruded into the longleaf growth at many places. The loblolly pine is abundant, and in tracts of virgin forest cuts of 12,000 to 15,000 board feet per acre were not uncommon. Early experts estimated that the loblolly growth covered an area of about seven thousand square miles and included all or parts of San Jacinto, Walker, Montgomery, Harris, Jefferson, Liberty, Orange, Hardin, Grimes, Newton, Jasper, and Chambers counties. In addition isolated pockets of so-called lost pines are found in Brazos, Fayette, Colorado, and Bastrop counties.[6] The loblolly has excellent reproduction char-

[3] William T. Chambers, "Pine Woods Region of Southeastern Texas," *Economic Geography* 10 (July, 1934): 303.

[4] Bray, *Forest Resources of Texas*, p. 21.

[5] Ibid., pp. 22–23; J. H. Foster, H. B. Krausz, and

George W. Johnson, *Forest Resources of Eastern Texas*, Texas A&M College, Department of Forestry Bulletin no. 5 (1917), p. 9.

[6] Bray, *Forest Resources of Texas*, pp. 19–21; J. H. Foster, H. B. Krausz, and A. H. Leidigh, *General Survey of Texas Woodlands, including a Study of the Commercial Possibilities of Mesquite*, Texas A&M College, Department of Forestry Bulletin no. 3 (1917), pp. 15–17.

acteristics, and much of the second-growth forest in the southern half of East Texas is of this type or of mixed loblolly and hardwood.

The shortleaf pine is found north of the long-leaf area and extends to the Red River. This region covers about thirty thousand square miles and was the first area opened to commercial lumbering because of the early arrival of railroads. In comparison with the longleaf and loblolly stands, the shortleaf forests are more open, more mixed with hardwoods, and more often interspersed with farm and grazing land. Shortleaf was found north of Nacogdoches and as far west as Hopkins and Anderson counties. In comparison to the region farther south, the landholdings were smaller, and the yield of board feet per acre considerably lower than that of the longleaf or loblolly areas. Lumbering and "tree farming" were often secondary instead of primary occupations of the landowners in the shortleaf belt. This was especially true as one moved north and west toward the blackland prairies.[7]

To the visitor first entering the region the towering pine forest was almost overpowering. Travelers often described the magnificent pines (probably longleaf) soaring 100 to 150 feet in the air with bases 4 or 5 feet in diameter. The forest floor under the great longleaf trees was clean, and the forest was described as parklike. Here the combination of sandy soil and woods fires had eliminated most competing growth, and the traveler walked or rode through the forest without difficulty. As he proceeded north and west from the Sabine and the Neches, the longleaf gave way to other varieties and to scrub hardwood mixed with pine. To all the travelers the great forest of pine trees with their majestic trunks pointing skyward, often 50 or 60 feet to the lowest limb, was a spectacular sight to be recorded and described again and again.

In 1828, Stephen F. Austin characterized the East Texas country as abounding in good pine

timber, with some cypress, cedar, and other varieties. He predicted that "among the staple industries of Texas would be lumber."[8] Three years later he described the region somewhat more fully:

The country on the Sabine, Neches, Trinity, and San Jacinto is heavily timbered and wooded with thick groves of good pine, cypress, oak, ash, and other timber. The level region extends about seventy miles from the coast in this section of Texas (east of the San Jacinto). . . . The thickly wooded lands continue quite to Red River north of the heads of the Sabine and the Neches, and pretty high on the Trinity.

The whole of this eastern and wooded region is very abundantly supplied with living streams of pure water, which afford many favorable sites for saw and other mills, either water or steam. The lumber business from this quarter will be very valuable so soon as mills are put in extensive operation.[9]

A few years later the anonymous author of *Texas in 1837* described the land surrounding Buffalo Bayou in these terms:

The land on the banks at this season of the year is generally dry, and the country back has the appearance of a plain. The soil is generally sandy, which will forever make the land of little value except for the timber, which exists in great abundance, especially the pine.[10]

The Frenchmen Frédéric Leclerc and Eugène Maissin and the Englishman William Bollaert, all of whom visited Texas during the period of the Republic, commented on the great size and extent of the pine forests. Bollaert, obviously impressed, wrote that "the pine grows in this country to great height, say 150 to 200 ft." He thought that there was more valuable timber on one estate of less than one league that he had visited in East Texas than on 500,000 acres of United States government land in Florida.[11]

[7] Bray, *Forest Resources of Texas*, pp. 24–25; G. Loyd Collier, "The Evolving East Texas Woodland" (Ph.D. diss., University of Nebraska, Lincoln, 1964), pp. 40–101.

[8] Eugene C. Barker, ed., "Stephen F. Austin's Descriptions of Texas," *Southwestern Historical Quarterly* 28 (October, 1928): 98–104.

[9] Ibid., p. 106.

[10] Andrew Forest Muir, ed., *Texas in 1837: An Anonymous Contemporary Narrative*, p. 22.

[11] Frédéric Leclerc, *Texas and Its Revolution*, trans. and ed.

Almost half a century later Frank H. Taylor, writing in *Harper's Monthly*, described the "Piney Woods," which were still largely a great virgin forest. "The wooded country of East Texas," he wrote, "yields a rich variety of useful woods, yellow pine, cypress, red and white oaks, liveoak, hickory, pecan, and cedar predominating. The Trinity, Sabine, Neches, Angelina, San Jacinto, and other rivers will afford rafting facilities at times."[12]

George L. Crocket, Episcopal clergyman and local historian, spent a lifetime in East Texas and witnessed the entire course of bonanza lumbering from 1880 to 1930. He described the virgin forests in eloquent terms:

The traveler entering Texas through the gateway El Camino Real found himself at once in a country of great natural beauty and fertility in the red land belt which runs east and west along the course of the highway. . . . In its virgin state there was little or no undergrowth save along the watercourses, but the trees rose in stately grandeur from a luxuriant carpet of the finest green. The wild life was abundant. Herds of deer might be seen feeding along the watercourses or on the hillsides; wild turkeys were plentiful; the forests were full of squirrels; coveys of game birds were abundant; the trees were vocal with feathered songsters; and in the winter the streams and ponds were the resort of innumerable water fowls.[13]

The geographer described the region in a more scientific if more prosaic manner. According to one leading Texas geographer, the East Texas timber belt had developed "upon a vast expanse of sand, sandy shale, and shale with some greens and lignite deposits."[14] The great mass of the bedrock is of the Tertiary period but extends westward upon Quaternary-period materials in the Houston–Beaumont area. Everywhere the terrain is level to gently rolling, and rainfall is abundant. The timber belt can be divided into subregions, the "pinewoods region" and the "East Texas agricultural region." The latter region is also forested, though not as heavily as the former, but also has considerable agricultural interests as well as forest industries.[15]

These subregions can be divided again roughly along the line of El Camino Real, the old San Antonio Road (Natchitoches, Louisiana, to Nacogdoches, to San Antonio). This division is marked also by a redland section extending westward from the Sabine on a narrow frontage and including some parts of five or six counties. Here the land is characterized by relatively fine-textured, red-colored soils, which present a marked contrast to the loose, light-colored sandy soils on the north and south. The bright-red color is attributed to a considerable quantity of iron oxide in the soil, "and some ridges and hills are capped or flaked by layers of low grade iron ore."[16] This redlands soil is more fertile than the sandy soils and contains more organic matter. Hence it is more productive, and much of it has been cleared for crops or pasture. Geologically, its presence is explained by substrata of "calcareous greensand or greenish clay containing glauconite, shell, high percentages of iron, and in places thin beds of limestone."[17]

Other redlands occur in East Texas, characterized by the same properties displayed by the wedge-shaped redlands zone in central East Texas along the Camino Real. Although not fertile in comparison with the rich blacklands of Central Texas, the redlands are more productive than the sandy soils, and the farmers there are more prosperous and progressive.

James L. Shepherd III, p. 24; Eugène Maissin, *The French in Mexico and Texas*, p. 179; William Bollaert, *William Bollaert's Texas*, ed. W. Eugene Hollon and Ruth Lapham Butler, pp. 266, 307.

[12] Frank H. Taylor, "Through Texas," *Harper's New Monthly Magazine* 59 (October, 1879): 706.

[13] George L. Crocket, *Two Centuries in East Texas: A History of San Augustine County and Surrounding Territory from 1865 to the Present Time*, pp. 80–81.

[14] William T. Chambers, "Geographic Regions of Texas," *Texas Geographic Magazine*, Spring, 1948, p. 8.

[15] Ibid. For a more recent description of the region see Edwin J. Fescue, "East Texas: A Timbered Empire," *Journal of the Graduate Research Center, Southern Methodist University* 28, no. 1 (April, 1960): 1–57.

[16] William T. Chambers, "The Redlands of Central Eastern Texas," *Texas Geographic Magazine*, Autumn, 1941, pp. 1–14.

[17] Ibid.

The older population centers of this belt are built upon the redlands.[18]

The geographic unity of the East Texas region is, however, inescapable. Despite the segregation of the subspecies, the whole area is essentially one great pine forest. Throughout the region the annual rainfall exceeds forty inches and approximates fifty inches in the areas of greatest timber density. The unifying industry is lumbering and the production of lumber products.[19]

Prominent among the geographical features of the East Texas region are its rivers. Generally speaking, the rivers of Texas run northwest to southeast, flowing into the Gulf of Mexico. At the eastern border of the state, the East Texas region is drained by the Sabine; the Neches and its major tributary, the Angelina; the Trinity; and the somewhat shorter San Jacinto. West of the Piney Woods runs the Brazos, and forming the northern boundary of the region and the state is the Red River. All these rivers played a significant part in the lives of the people both before and during the development of the lumber industry.

By the time of the Civil War a number of towns had grown up on the rivers and served as shipping ports for the surrounding country. The East Texas region had developed a kind of subsistence agriculture that featured relatively small farms in the midst of great timbered landholdings. The cleared and open spaces, especially those in the redlands, were valued, but the timbered sections were considered of little worth, almost a liability. Despite the primitive state of agriculture, the farmer produced a surplus of commodities, principally cotton and beef and pork products, and transported them to market. He had a choice of routes: he could haul his products by wagon to the population centers or send them downriver by boat or barge. Wherever feasible, most farmers preferred the rivers.

Much could be written about the Sabine region. It was the country of the "neutral ground," in the eighteenth century a dividing strip between the domains of Spain and France and in the early nineteenth century a vague no-man's land beyond the rather indistinct claims made by the United States on the basis of the purchase of the Louisiana Territory. It was a haven for refugees and the base for such filibustering expeditions as those led by Augustus Magee, Dr. James Long, and Philip Nolan. A number of places along the Sabine were points of entry during the Mexican period, and they were the goals of Texas settlers fleeing from the wrath of Santa Anna in 1836. It was along the Sabine that American General Edmund Pendleton Gaines deployed his troops to allow Houston and the Texas army to pass through to safety should a haven become necessary and to bar the further advance of Santa Anna. These events and many more were part of the history of the Sabine by the time Texas became a state.

Along the course of the Sabine were settlements (many of which have now disappeared) that were regular ports of call for steamboats in the years before the Civil War. These ports bore such names as Sabine City, Point Young, Green's Bluff, Princeton, Salem, Belgrade, New Columbia, Sabine Town, Patterson's Ferry, Hamilton, and Logan's Port. For a time they did a thriving business shipping cotton, hides, and beef to the coast and returning a variety of manufactured goods. Lumber, staves, and shingles were also among the goods exported, though these products often went downstream by barge or raft rather than steamboat. As early as 1838 the steamboat *Big Ben* was reported to have reached Belgrade. By the 1850s such river captains as William Wiess (or Weiss) regularly ran steamboats to Logan's Port. It was estimated that about forty boats were operating at that time on the Sabine and the Neches.[20]

[18] Ibid.
[19] William T. Chambers, "Pine Woods Region of Southeastern Texas," *Economic Geography* 10 (July, 1934): 302–18; William T. Chambers, "Divisions of the Pine Forest Belt of East Texas," *Economic Geography* 6 (January, 1930): 94–103.

[20] *Texas Almanac*, 1911, p. 126; Cleo F. Evans, "Transportation in Early Texas" (master's thesis, St. Mary's University, San Antonio, 1940), p. 76; Bernice Lockhart, "Navigating Texas Rivers, 1821–1900" (master's thesis, St. Mary's University, San Antonio, 1949).

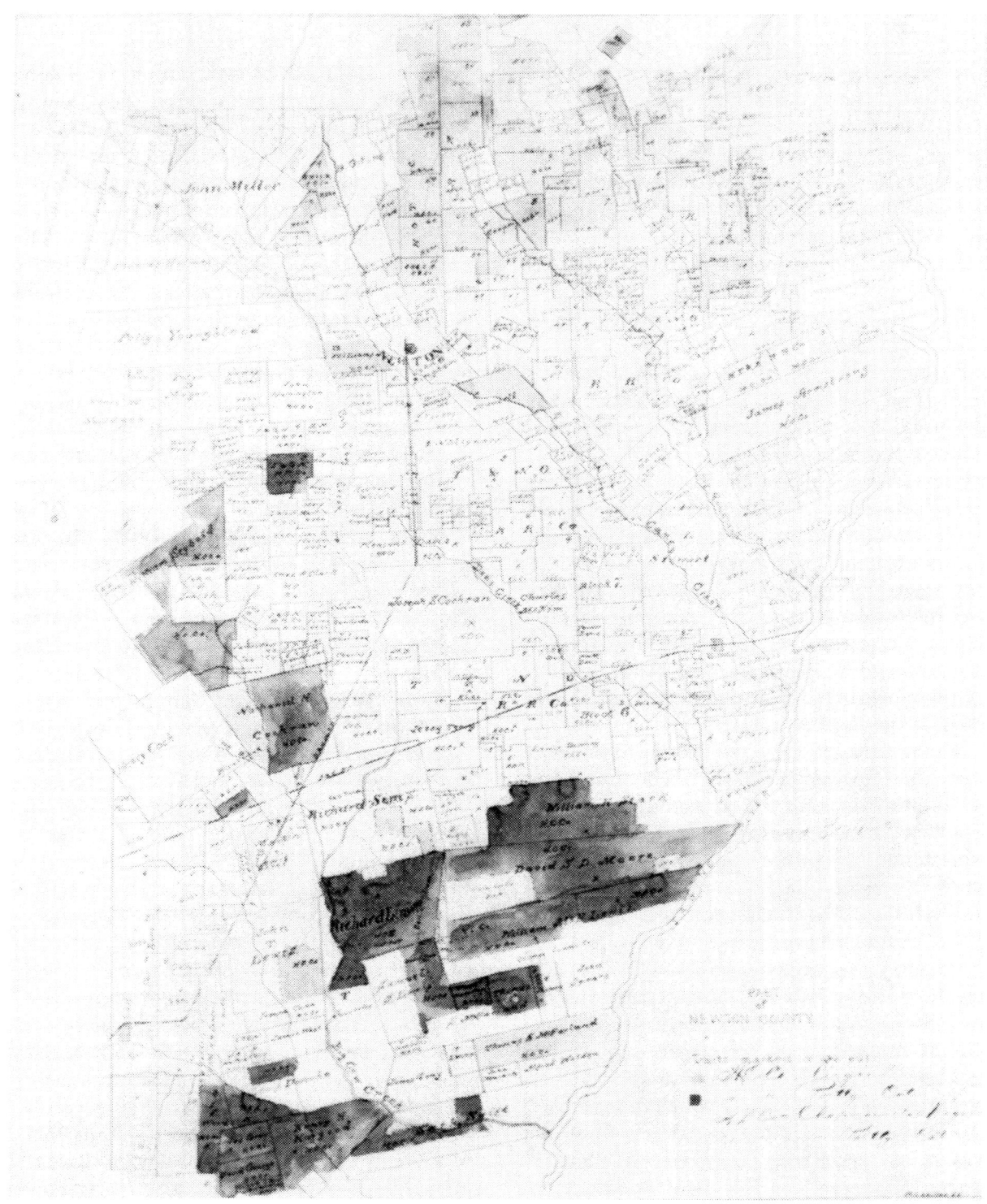

Land and timber holdings in East Texas. *Authors'
collections.*

George L. Crocket described activities on the Sabine during the antebellum period:

Freight . . . frequently came up the Sabine River and was unloaded at Sabine Town, Pendleton, Hamilton and other points along the river. General John G. Berry owned a steamboat called the "Big Ben" which made regular trips up the river, delivering goods at the landing points and taking back cotton on the return trip. This steamboat traffic on the Sabine was opened up as early as 1838, and Thomas S. McFarland in his diary tells of watching the first boat steam up the river past his newly laid off town of Belgrade.[21]

Considerable traffic also developed on the Neches and the Angelina, and shipping ports were established before the Civil War for the convenience of the farmers and ranchers. In addition to Beaumont the ports included Town Bluff, Teran, and Rockland on the Neches and Bevilport, Marion, Pattonia, and McNeill's Landing on the Angelina. In periods of high water boats and barges navigated as far as Rockland, Pattonia, and Marion, but during the low water of summer the head of navigation was no higher than Wiess Bluff, about twenty-eight miles above Beaumont.[22]

The Trinity River, which empties into Galveston Bay, provided an even greater waterway for inland traffic. Beginning with Anahuác, at its mouth, river towns on the Trinity included New Boston, Liberty, Grand Cane, Smithfield, Swarthout, Geneva, Sevastopol, Trinity, Cincinnati, Alabama, Pine Bluff, Buffalo, Providence, and (for the promoters and optimists) Dallas. No fewer than twelve steamboats were reported operating regularly on the Trinity before 1860, headed by the *Scioto Belle*, which went upriver as early as 1836.[23] The traffic on the Trinity was similar to that on the Neches and the Sabine: foodstuffs, salt, and household goods went upriver; cotton, hides, beef, staves, shingles, and some sawed lumber went downstream. Unfortunately, like the other East Texas rivers, the Trinity was navigable only one-quarter as far in the summer and fall as during the high water of the winter and spring. Nevertheless, a number of these settlements developed into considerable shipping ports and did a thriving river traffic. Especially on the Trinity the hope was always strong that funds from the federal government or the state would improve the river so that it would be navigable to Dallas County.[24]

Among other rivers that drained East Texas, the San Jacinto and Buffalo Bayou provided some navigation to the vicinity of Houston, which with improvement was later developed into the Houston ship channel. In describing this body of water, the anonymous visitor to Texas in 1837 wrote:

The banks are high and lined with the cypress knee, a bulbous root which shoots up along the edge of the water. In passing over this singular body of water, which is confined with but few exceptions, where small lakes or tributaries put in, to precipitous banks on either sides, covered with massive timber, consisting besides the cypress, of the stateliest pine, the oak, the wild peach, somber shade over its dark and sluggish waters, one cannot but imagine that he is drawing to the abode of some evil spirit, whose genius is stamped upon everything that meets the eye.[25]

On the north, Red River, Caddo Lake, and Big Cypress Bayou made Jefferson for a time a thriving commercial center and a shipping point for East Texas products. Because of a river-clogging raft above Caddo Lake in Red River, Jefferson was the head of navigation, and in the years immediately before and after the Civil War shippers sent lumber, cotton, hides, beef, and other commodities to New Orleans by way of Red River. During this period Jefferson was the largest town in Northeast Texas, claiming a population of twelve

[21] Crocket, *Two Centuries in East Texas*, p. 222.
[22] *Texas Almanac, 1911*, p. 126; Lockhart, "Navigating Texas Rivers," pp. 76, 102.
[23] Robert E. Mills, "Navigation of the Trinity River" (master's thesis, Sam Houston State Teachers College, Huntsville, 1943); map, "Southeast Texas, 1861," Stephen F. Austin State University Library, Nacogdoches.
[24] Mills, "Navigation of the Trinity River," p. 4; *Texas Almanac, 1870*, p. 191.
[25] Muir, ed., *Texas in 1837*, p. 22.

thousand in 1870. In time, however, the re-
moval of the raft lowered the water level of Big
Cypress Bayou, and boats could no longer
make the trip to Jefferson. As a result the town
declined and ceased to be important in the
transportation picture of the Piney Woods.[26]

Texas rivers were notoriously unreliable. In
the summer months they became so low that
even smaller boats could not navigate them for
any considerable distance. In the winter and
spring the water level rose suddenly, sometimes overnight, threatening to carry away
barges, boats, and their cargoes. Sandbars
shifted unexpectedly, and at all seasons logs,
snags, and sleepers (submerged sawyers) were
constant threats to boats. The verdict of the
anonymous visitor to Texas in 1837 was probably accurate:

It cannot be said that there is one good navigable
stream in all Texas unless we are disposed to con-
sider as such those which may be navigated during
the flood season of the year. But at such time they
cannot be called good. Steamboats of a very light
draft may run up a few of the streams of this country during the summer for a short distance from the
coast, but even then the navigation is dangerous on
account of the shallows and other obstructions. To
say the most, the rivers of Texas of the first class
will not compare with those of the United States
of the same grade in point of commercial advantages, leaving the Mississippi entirely out of the
question.[27]

What of the people who lived in the Piney
Woods in the early days? Added to a scattered
base of long-term Spanish-Mexican residents
was an Anglo-American population, most
members of which had migrated to Texas after
1820 from other southern states. According to
a sampling of the 1860 census, 20 percent of
the settlers came from Alabama, 16 percent
from Tennessee, and 15 percent from Mississippi. Next in order of numbers were settlers
from Arkansas, Georgia, Louisiana, Missouri,

and Kentucky. No single northern state contributed as much as 3 percent of the total.
There was a scattering of immigrants from
western Europe. In 1860 east of the Trinity
there were about eighty thousand black slaves,
composing 31 percent of the total population.
Thus East Texas was distinctly southern and
resembled closely the population pattern of
the older southern states on the east.[28]

The economy of the region was based on agriculture—subsistence farming in which cotton was the principal money crop. The climate
was warm, the winters were mild, and corn
and other basic crops grew with a minimum of
labor. Most farmers owned herds of cattle,
which grazed in the forests, and an undeter-
mined number of hogs, which ran wild. The
wealthier farmers or planters owned slaves,
held large acreages of land (mostly covered
with timber), and raised cattle, hogs, and
horses but usually had little cash. They laboriously hauled cotton by ox-drawn wagon
to Shreveport or Natchitoches, where it was
loaded on steamboats or barges plying the Red
River. Alternatively they hauled the cotton
to one of the shipping ports on the Sabine,
Neches, or Trinity and hoped for enough rain
in the early fall to make the river navigable.[29]
Historian Crocket described early-day freight
hauling in East Texas:

The goods were shipped to the nearest landing on
Red River, usually to Natchitoches, and hauled
thence on ox-wagons. A heavy wagon with three
yoke of oxen would carry about three thousand
pounds, and would take six bales of cotton in and
bring a load of freight back, for which they were
paid at the rate of one dollar per hundred pounds.
Three or four weeks were usually consumed on the
trip, so that the wagoner had to take with him supplies for himself and his team. When possible a
number of wagons would go together so that they
could help each other through the bog and mud of

---

[26] Mrs. Arch McKay and Mrs. H. A. Spellings, A History of
Jefferson, 5th ed., pp. 7–8; Winnie Mims Deal, Jefferson,
Texas: Queen of the Cypress.

[27] Muir, ed., Texas in 1837, p. 119.

[28] Barnes F. Lathrop, Migration into East Texas, 1835–1860,
p. 55.

[29] Crocket, Two Centuries in East Texas, pp. 233–55; Virgie Scurlock, "Ante-Bellum Nacogdoches, 1846–1861" (master's thesis, Stephen F. Austin State College, Nacogdoches,
1954), 2:22–25, 28.

Hardwood bottomland forest. *Courtesy of Stephen F. Austin State University, Forest History Collections, Thompson photo albums.*

the creek bottoms. They camped together at night and prepared their meals, consisting of corn pone and fried bacon, with strong coffee, and if there happened to be a barrel of whiskey in the load, a rifle ball fired through the end would furnish what liquor they wanted, "for medicinal purposes" as they would explain to the owner of the barrel upon arrival.[30]

The roads were miserable and during the winter virtually impassable. Although once denoted El Camino Real (the King's Highway), the San Antonio Road had deteriorated in many places to little more than a double-packhorse trail marked with mudholes and stumps. Other roads, though well marked on maps, were even worse. For example, in the 1870s, during dry weather it took two days to deliver mail to Nacogdoches from Crockett, 60 miles away, the nearest point on the railroad. In the winter it took five days to a week.[31] In an effort to improve the roads, the state Legislature in 1850 chartered the Angelina Bridge and Turnpike Company to build a bridge across the Angelina River and construct and maintain a north–south turnpike. The project was completed, and the company was authorized to charge a toll of 60 cents for a four-wheeled carriage, 25 cents for a wagon, 10 cents for a man and horse, 5 cents for a man on foot, and 2 cents a head for cattle. Despite the company's efforts the roads did not remain in good condition, and by the end of the Civil War the company did nothing more than collect tolls at the bridge.[32]

The isolation, poor transportation, mild climate, and relatively easy living appeared to have an ill effect on the people of the Piney Woods, many of whom travelers saw as slovenly, careless, and unambitious. The travelers commented unfavorably on these characteristics. When Mrs. Mary A. Maverick, of San Antonio, traveled through East Texas in 1838, she described her journey as a trip in "occasional rain and much mud, where the country was

poor and sparsely settled and provisions for man and beast scarce."[33] Architect Frederick Law Olmsted described a young Texas planter whose father, a mechanic, had migrated to East Texas from the North and left his son some property in land and cattle. "His house," said Olmsted, "was more comfortless than nine-tenths of the stables in the north. He was without care, thoughtless, content, with an unoccupied mind. He took no newspaper. He read nothing."[34] Olmsted quoted a similar comment by a northern woman that there was little comfort in East Texas because "the people would not take the trouble to get anything."[35]

Except for Galveston, Houston, and Jefferson, the towns were little more than villages. In 1784, Nacogdoches, perhaps the oldest town in the region, was reported to have 399 inhabitants. According to the federal census of 1850 and later, the census of 1880, the population of the town still numbered about 400. The other towns were similar: small, somewhat grubby settlements in which most of the people merely "got along." Yet there were also people of culture and broad interests in the East Texas towns. Both San Augustine and Nacogdoches boasted "universities" before the Civil War, and Moscow, in Polk County, was described as a center of education. Such men as Adolphus Sterne, Dr. Robert Irion, and Frost Thorn, of Nacogdoches, and Dr. Joseph Rowe, S. W. Blount, and James P. Henderson, of San Augustine, were persons of polish and charm as well as substantial men of affairs. Each of the towns supported a number of churches, and most had flourishing local chapters of Masons.[36]

Olmsted described in some detail the journey he made through East Texas during the winter of 1853. Crossing from Louisiana, he noted "the heavy clay soil, stained by an oxide

---

[30] Crocket, *Two Centuries in East Texas*, p. 222.
[31] Taylor, "Through Texas," p. 705.
[32] Scurlock, "Ante-Bellum Nacogdoches," pp. 25–26.

[33] J. Frank Dobie, *The Flavor of Texas*, p. 174.
[34] Frederick Law Olmsted, *The Cotton Kingdom*, p. 304.
[35] Ibid., p. 301.
[36] Scurlock, "Ante-Bellum Nacogdoches," 1:3–4; Crocket, *Two Centuries in East Texas*, pp. 233–54, 315.

of iron to a uniform brick red. It makes most disagreeable roads, sticking close, and giving an indelible stain to every article that touches it." Olmsted found the people the "most sturdily inquisitive I ever saw. Nothing staggers them, and we found our account in making a clean breast of it as soon as they approached."[37]

According to Olmsted, "San Augustine made no very charming impression as we entered nor did we find any striking improvement on longer acquaintance." He counted fifty or sixty houses and a half a dozen shops and stores. Most of the stores fronted on what he called a "central square acre of neglected mud." In some contrast, he described Nacogdoches as a considerable town, with neatly painted houses displaying the "first exterior sign of civilization since the Red River."[38] About the same time Mayor John Forbes described Nacogdoches as "far behind most other towns of more recent creation, in population, improvement, and enterprise."[39] Evidently Forbes strongly disliked the Spanish influence of the town, while Olmsted found it rather attractive.

The eating and reading habits of East Texas residents drew comments from many travelers. Olmsted described an evening meal—a rather typical one—as follows: "On our supper table was nothing else than the eternal fry, pone and coffee. Butter, of a dreadful odor, was here added by exception. Wheat flour they never used. It was too much trouble."[40]

Solomon Wright recalled the standard fare of early-day East Texans: "Their diet would by no means please the stomach of an epicure. Corn-bread, bacon and potatoes, with an oc-

Logger Peter Doucette measures his girth against that of a single virgin pine, ca. 1908. *Courtesy of Stephen F. Austin State University, Forest History Collections.*

casional treat of venison, give them perfect satisfaction." He described the fare he received as a potluck guest on one journey:

After a hard day's ride I stopped at a house near the road for supper and shelter for the night. About fifteen minutes after my arrival my host announced supper was ready. I cast my eyes over the anticipated meal. My digestive organs, after the inspection of the supper spread before me, rebelled and contracted. The following is the bill of fare complete: Cornbread, very fat bacon, and clabber. As I am not fond of clabber, I did not eat it. My host called his daughter and said: "Emma Jane, bring this man some water." My heart felt sick within me to think I could not get a cup of coffee.[41]

Despite the libraries of books in the homes of a number of leading citizens, such as Thorn, Sterne, and Blount, the nonreading

[37] Frederick Law Olmsted, *A Journey through Texas*, pp. 67–68.

[38] Ibid., pp. 68–79.

[39] Scurlock, "Ante-Bellum Nacogdoches," p. 21.

[40] Olmsted, *The Cotton Kingdom*, p. 301. Wheat flour, of course, would have had to be imported, since the combination of heat, humidity, and midsummer rains would have prevented successful wheat growing in the region. Only wealthy southerners could afford to import wheat flour. In spite of the tone of his comment, Olmsted probably knew this.

[41] Solomon Alexander Wright, *My Rambles as an East Texas Cowboy, Hunter, Fisherman, Tie-Cutter.*

habits of the average East Texan appalled Olmsted. When he had completed his trip, he commented that "in the whole journey through East Texas we did not see one of the inhabitants look into a newspaper or a book, although we spent days in houses where men were lounging about the fire without occupation."[42]

The nonreading habit can doubtless be explained by the high rate of illiteracy among the adults in the region. According to the manuscript census of 1860 for Nacogdoches County, which was not one of the more "backward" counties in the area, 227 adults of 1,090 households, "including roughly equal numbers of whites and Mexicans," were listed as illiterate.[43] Doubtless twice that number read so painfully and slowly as to be functionally illiterate, unable to comprehend any but the simplest printed matter.

East Texas was also a violent region, its citizens much given to settling disputes directly with rifle, pistol, or knife without waiting for the law to take its course. So pervasive was the feuding tendency that much of the history of central East Texas before the Civil War centered around the war between the "Regulators" and the "Moderators" and the efforts of the authorities to put down the violence. The conflict seemed to infect every aspect of life. Rivalry between the Methodist and Presbyterian "universities" in San Augustine became so heated that the president of the Presbyterian school was shot down on the street, in an act that hastened the demise of both institutions.[44] Two decades later a traveler observed that the tendency to violence had not diminished. He noted that, although about three hundred murders had been committed in Texas during the previous twelve months, only eleven executions had been recorded.[45]

Both time and space are important in delimiting a region.[46] In many respects central East Texas in the 1870s was looking backward and was less progressive than it had been in the 1850s. East Texans associated free schools with the hated Reconstruction policies imposed during the administration of Governor Edmund J. Davis, and, although there had been many educational ventures in East Texas before the Civil War, many generations would pass before tax-supported compulsory public education was fully accepted there.[47] Thus it was not surprising that most East Texas counties lagged far behind the rest of the state in education or that the more sophisticated centers, Dallas, Galveston, and Houston, poked fun at the "backwoods" character of the people and spoke of East Texas as the "Barefoot Nation."[48]

Immigrants to any new country, sharing the same cultural and religious background, tend to congregate together.[49] They bring with them their habits, mores, and manner of speech. Isolation and intermarriage perpetuate these patterns long after they have died out in more mobile and populous areas. Hence both nineteenth- and twentieth-century travelers have noted expressions peculiar to the East Texas region, brought there from the uplands of Alabama and Mississippi and the highlands of Tennessee. People would "porch" eggs and "warsh" clothes and say "far" for "fair," "bar" for "bear," and "Jurden" for "Jordan." A person or thing of little worth was "no account," and a farm was rated to "make ten to fifteen gallons of corn per acre." Expressions such as "done gone," "I wonders," and "I

[42] Olmsted, The Cotton Kingdom, p. 301.

[43] Scurlock, "Ante-Bellum Nacogdoches," 2:278–544.

[44] Crocket, Two Centuries in East Texas, p. 306; Olmsted, Journey through Texas, p. 69.

[45] Taylor, "Through Texas," p. 709.

[46] John C. Appel, "Regionalism and American History," Social Education 13 (November, 1949):319–24.

[47] Thomas C. Richardson, East Texas: Its History and Its Makers, p. 481.

[48] Lewis Nordyke, The Truth about Texas, p. 111.

[49] This is demonstrated by the pattern of immigration into East Texas counties as shown by the 1860 census. For example, the great influx of immigrants from Alabama and Mississippi settled in the central East Texas counties, while Tennessee immigrants mostly settled in Nacogdoches County and north. Settlers from Kentucky chose the more northern counties, while those from Louisiana settled in the southeastern counties. See Lathrop, Migration into East Texas, 1835–1860.

A grown man is dwarfed by the old-growth long-leaf pine forest. *Courtesy of Stephen F. Austin State University, Forest History Collections.*

never sawed" were common. Indeed, in remote areas some of these expressions linger to the present.[50]

One of the most unusual and characteristic aspects of the East Texas region was the survival of "sacred-harp" singing. According to most historians, the custom was brought to the Piney Woods by way of Tennessee and Alabama from seventeenth-century England. The music has the cadence of the ancient English ballads and stems from the Elizabethan period. Instead of the usual score of eight notes, one form of sacred-harp music has only three (fa, sol, and la). In the printed music the notes are designated by distinctive shapes and hence are called "shaped notes." For more than a hundred years sacred-harp music remained popular in the Piney Woods, and great throngs gathered to sing in churches, halls, and the open air. Like other customs, this custom too has survived to the present.[51]

In the decade after the Civil War the great Texas pine forest stood largely as it had for centuries: isolated, magnificent, and virgin. The settlers in the region had not molested the forest except to clear a fraction for cotton culture and for buildings. Land transportation was difficult, and the rivers, though numerous, were unreliable for travel. The population, made up largely of southern Anglo-Saxon stock, engaged in an easygoing form of subsistence farming, ranching, and cotton culture on relatively poor land ill-suited for the purpose. Intermingled with the Anglo and Mexican population were almost 100,000 former slaves to whom fell most of the drudgery and hard work. Although there were in every town families of property, education, and enterprise, there were very few that did not view the forests as obstacles of little worth. Timbered land was of less value than cleared land.

It remained for newcomers, experienced lumbermen, and adventurous entrepreneurs from the older states of the South, North, and East and from Europe to envision the great potential value of the Texas Piney Woods and provide the capital, expertise, and energy to exploit them.

---

[50] Olmsted, *Journey through Texas*, pp. 77–79; Nordyke, *The Truth about Texas*, p. 111; Richardson, *East Texas: Its History and Its Makers*, p. 479.

[51] Nordyke, *The Truth about Texas*, p. 118. See also Francis E. Abernethy, *Tales from the Big Thicket*, pp. 200–14.

CHAPTER 2

The First Big Mill

Through the forest to great riches and
power . . .—preface

FROM the time of the earliest travels and explorations of whites in East Texas, the great pine forests attracted their attention and served as a source of fuel and timber. Perhaps the first loggers were the Spanish priests and monks who, in 1716 and 1717, established missions to the Indians of East Texas. These missions were later abandoned, the buildings decayed and disappeared, and the forests reoccupied the sites so completely that even their locations are uncertain.[1]

A century later, when Stephen F. Austin first came to Texas, there were some primitive hand-sawing operations (probably by pit sawyers—see below) at several places along the rivers and near the coast. Apparently some water-powered mills began operations on an irregular basis at various locations during the decade or so before 1830. There is a record of a water-powered mill at San Augustine as early as 1819 and of another mill that operated with a horse-powered treadmill in 1825.[2] Near Nacogdoches, sometime before 1829, Peter Ellis Bean, a veteran adventurer and former Mexican-government official, built a water-

powered combined sawmill and gristmill. He also ran a lumberyard in town, which, to judge from surviving letters and documents, maintained a large inventory of building materials.[3] Similar sawmills were in operation in Cherokee County and doubtless elsewhere in the region during the same period. All of these were small, water-powered sash mills (see below), which crudely and laboriously produced lumber at a rate of no more than 500 to 2,000 board feet a day. The power to operate such mills was obtained by damming local creeks.[4]

Steam-powered sawmills made their appearance in East Texas during the early 1830s. Recognizing the need and the opportunity, Mary Austin Holley, a cousin of Stephen F. Austin, predicted that, with no want of timber and a rapidly increasing demand for lumber, "fortunes may be realized by good sawyers, with slight capital."[5] At about the same time, John Richardson Harris, the founder of Harrisburg, decided to establish a mill and journeyed to New Orleans to purchase sawmill machinery. He died of fever while in the Crescent City, and his brothers, William P. and Da-

Material in this chapter substantially appeared in *Southwestern Historical Quarterly* 86, no. 1 (July, 1982): 1–30, and is used with its permission.

[1] Rupert N. Richardson, *Texas: The Lone Star State*, 2d ed., pp. 17–25.

[2] Green Payton, *The Face of Texas*, p. 85; *Texas Almanac, 1958–1959*, pp. 316–17. A historical marker on State Highway 21, San Augustine, records the location of the sawmill built in 1819 on Ironsa Creek, one-quarter mile north.

[3] R. B. Blake Papers, Nacogdoches Archives, Stephen F. Austin State University Library, Nacogdoches, 3:142–49; hereafter cited as SFA Library. In 1832, Bean and Frost Thorn purchased a league of land together with a "sawmill and gristmill" on La Nana Creek from José Mora. Bennett Lay, *The Lives of Ellis P. Bean*, pp. 153, 198.

[4] Hattie Joplin Roach, *A History of Cherokee County, Texas*, p. 87.

[5] Mary Austin Holley, *Texas*, pp. 71, 120.

Pit sawyers laboriously cutting one board at a time. *Courtesy of the artist, Bruce Lyndon Cunningham.*

Sash (up-and-down) saw, operated by water power, animals, or steam. *Courtesy of Forest History Society.*

vid Harris, with Robert Wilson, brought the machinery to Texas and set up a mill near Harrisburg before the end of 1830. This mill was one of the first, if not the first, of the steam sawmills in the Buffalo Bayou region. It was described as a small, steam-powered circular-saw outfit with a "make-shift Arkansas smoke kiln."[6] It operated for several years, producing lumber that the owners sold locally or shipped throughout Central Texas, sending some wagons as far as San Antonio. According to early settlers, General Santa Anna's troops burned the mill along with much of the town during the invasion of April, 1836.[7]

As the Republic grew after the revolution, the demand for lumber caused a number of sawmills to spring up along Buffalo Bayou and elsewhere in Southeast Texas. A visitor in 1837 observed that "lumber of all kinds was hard to be procured and was selling for seventy dollars per thousand."[8] A year later the Frenchman Eugène Maissin commented on the rapid building of new sawmills and described Houston as "a town made of planks sawed yesterday and scarcely dry."[9] Boats, barges, and rafts carried much of the manufactured lumber down Buffalo Bayou to Galveston Bay, where it was transshipped to its destination.

Another center of early sawmill operations was the one along the Sabine. As early as 1836, Robert Booth (or Boothe) constructed a small sash mill north of present-day Orange that was said to be capable of cutting 1,500 board feet of lumber a day.[10] In the 1840s, Robert Jackson

[6] Bill Doree, "Texas' First Steam-powered Saw Mill," *Gulf Coast Lumberman,* April, 1963, p. 13.

[7] J. H. Kuykendall, "Reminiscences of Early Texans," *Quarterly of the Texas State Historical Society* 7 (July, 1903): 29–64; Adele B. Looscan, "Harris County, 1822–1834," *Quarterly of the Texas State Historical Society* 18 (October, 1914):195–207; "Journal of Lewis Birdsall Harris, 1838–1842," *Quarterly of the Texas State Historical Society* 25 (January, 1922):185–97; "The Pioneer Harrises of Harris County, Texas," *Quarterly of the Texas State Historical Society* 31 (April, 1928):365–73.

[8] Andrew Forest Muir, ed., *Texas in 1837,* p. 29.

[9] Eugène Maissin, *The French in Mexico and Texas,* translated and edited by James L. Shepherd III, pp. 185–86.

[10] James Boyd, "Fifty Years in the Southern Pine Industry," *Southern Lumberman* 144 (December 15, 1931):59–67; Thomas C. Richardson, *East Texas: Its History and Its Makers,* 3:1173.

operated a sawmill at Turner's Ferry, on the Sabine, which was described as a sash mill powered by an engine and boilers salvaged from a steamboat that had been wrecked upstream. In 1847, after a spring flood partly destroyed his operations, Jackson moved his mill to Orange (then known as Green's Bluff). It was said to be the first steam sawmill in Orange.[11]

The most common type of saw used in these early mills was the sash saw, a single sawblade held rigidly in a frame that moved up and down with the saw. It cut only on the down-stroke, and the log was forced against the saw by a crude mechanical feeder. Essentially it was simply a mechanization of the old, hand-operated pit saw, which two men, one standing above the log and the other standing in a pit below it, pushed and pulled until they produced a board. The power activating the sash saw could be animal power, waterpower, or, finally, steam power. From the sash saw developed the muley saw, in which the sash and frame were lightened and the stroke speeded up, and the first gang saw, in which additional blades were fitted into the frame to increase production. The lumber produced by any of these methods usually totaled no more than 500 to 2,000 board feet a day, and the boards were rough and uneven. It is not surprising that sawmill operators sought more efficient methods of manufacturing lumber.[12]

It is uncertain when the circular saw was introduced into the South. It is known that Samuel Miller received a patent on a metal circular saw in England in 1777 and that a Benjamin Cummins (or Cummings), a blacksmith in New York State, hammered out a successful circular saw about 1820. Although mention

has been made of the presence of small circular saws in Texas in the early 1830s, they were not in general use until near the beginning of the Civil War. Many operators feared the risk of injury from the circulars when they were running at high speed and objected to the loss of one-fourth to one-half inch of the boards at every cut owing to the wide kerf (cutting groove) of the average circular saw. Some manufacturers of competing saws described the circular saws of this era as sawdust-making machines that produced some lumber as a by-product. Nevertheless, the greatly increased production promised by a continuous cutting edge prompted most operators to shift to the circular saw shortly before or soon after the war.[13]

In 1860 the United States census reported approximately two hundred sawmills in Texas with about twelve hundred employees. The industry manufactured and sold lumber to the value of $1.75 million annually, of which it was estimated that 70 percent was added by the manufacturing process. About three hundred men were employed in auxiliary occupations that depended on the lumber industry, such as construction, carpentry, and cabinet-making.[14] By this time coastal lumbermen had developed a considerable export trade, especially with other Gulf Coast cities, such as Matagorda (Texas), Matamoras, and Tampico (Mexico). Sawmill owners in the interior of East Texas served mainly local needs, having no easy access to the Gulf, since the cost of hauling lumber long distances was often prohibitive. All these mills were small, whether sash or small circular saw, and often were combined with a gristmill or a general-merchandise store.[15]

---

[11] Richardson, *East Texas*, 3:1173: Boyd, "Fifty Years in the Southern Pine Industry," pp. 59–67.

[12] F. H. Gilman, "History of the Development of Saw Mill and Woodworking Machinery," *Mississippi Valley Lumberman* 26 (February 1, 1895):60–61; Stanley F. Horn, *This Fascinating Lumber Business*, pp. 22–23; Ralph W. Andrews, *This Was Sawmilling*, p. 44. See also J. Richards, "A Treatise on the Construction and Operation of Woodworking Machines," *Journal of Forest History* 9, no. 4 (January, 1966):16–23.

[13] Horn, *This Fascinating Lumber Business*, pp. 136–37; Gilman, "History of the Development of Saw Mill and Woodworking Machinery," pp. 61–62.

[14] Vera Lea Dugas, "Texas Industry, 1860–1880," *Southwestern Historical Quarterly* 59 (October, 1955):54.

[15] Bernice Lockhart, "Navigating Texas Rivers, 1821–1900" (master's thesis, St. Mary's University, San Antonio, 1949), p. 76; *Nacogdoches News*, July 30, 1877; March 16, 1882; *American Lumberman*, September 26, 1908, pp. 86–87.

Like most other business ventures of the Confederacy, lumbering declined during the Civil War years. Many mills ceased operations because of manpower shortage or lack of replacement machinery. In Texas production lagged, and reports showed that lumber exports fell to one-fourth their prewar volume.[16] With the end of the war in 1865 came a revival of interest in the industry; local lumbermen repaired and built larger mills, and a number of new residents arrived, some with investment capital and previous experience in the lumber business. John Martin Thompson, who had been operating a small mill in northern Rusk County since 1852, built a new, larger mill after the war and increased his production capacity to 8,000 board feet a day. Thompson, with his brothers and sons, built a series of mills, each larger than the last, until they moved to Trinity County in 1881. In the new location they continued to expand, finally establishing themselves among the major lumbermen of the state.[17]

Among the newcomers to lumbering after the war was Alexander Gilmer. A native of Ireland who had previously lived in Georgia, Gilmer established himself at Orange before the war and engaged in shipping and shipbuilding. During the conflict he successfully ran the Union blockade several times. In 1866 he entered the sawmill business and built a mill on the Sabine River, above Orange. After several mill fires he moved to Orange, where he eventually owned several mills and became one of the prominent figures in the industry.[18] Another early entrepreneur in the lumber business was Judge David R. Wingate. A former millowner on the Pearl River in southern Mississippi, Wingate moved to Texas just before the Civil War and engaged in a sawmill venture on the Sabine in Newton County. After the war he too moved to Orange and in 1874 built a new steam mill with a circular-saw rig that had an estimated capacity of 20,000 to 30,000 board feet a day. That production made Wingate, along with Gilmer, one of the larger lumber operators in East Texas. A sawmill of that size moved the owner from the "small-mill" category into the moderate, or "middle-size," class.[19] Among others who entered or reentered the lumber business after the war was Fritz Amsler, of Montgomery County, who also installed a planing mill. The Gibbs brothers began operation of the Palmetto Lumber Company in San Jacinto County in 1874. In Beaumont, W. A. Fletcher became a partner in the Long Lumber Company in 1870 and in 1876 helped organize the Beaumont Lumber Company, which grew to be an important force in the industrial life of that city.[20] All these lumbermen, Amsler, the Gibbs brothers, and Fletcher, became prominent lumbermen in the Southwest and by 1876 controlled sawmill operations that averaged 25,000 to 40,000 board feet of lumber a day. Like Gilmer and Wingate they had moved beyond the "small-mill" stage of operations.

Thus by the mid-1870s the lumber industry in Texas had largely recovered from the war and the prolonged depression that followed and was looking to the future. Yet in comparison with the production of the leading lumber-producing states, such as Michigan, Wisconsin, and Minnesota, Texas production was small; the great pine forests of East Texas were still virtually untouched. In 1870 the total cut in Texas was less than 100 million board feet, and ten years later it had only reached 300 million board feet.[21]

[16] Texas Almanac, 1867, p. 125.

[17] American Lumberman, September 26, 1908, pp. 67–150.

[18] Walter Prescott Webb, ed., Handbook of Texas, 1:691; Ruth A. Allen, East Texas Lumber Workers, pp. 16–18.

[19] Nollie Hickman, Mississippi Harvest: Lumbering in the Longleaf Pine Belt, 1840–1915, pp. 18–19; Allen, East Texas Lumber Workers, p. 142; Burke's Texas Almanac, 1979, pp. 31–32.

[20] Dugas, "Texas Industry," pp. 161–62; Ruth Hansbro, "History of San Jacinto County, 1870–1940" (master's thesis, Sam Houston State University, Huntsville, 1940), p. 77; Boyd, "Fifty Years in the Southern Pine Industry," pp. 59–67; Florence Stratton, The Story of Beaumont, pp. 135–36.

[21] Texas Almanac, 1958–1959, p. 190. For comparisons see Henry B. Steer, Lumber Production in the United States, 1799–1946.

A number of serious problems faced the fledgling Texas lumber industry. Not the least of these was lack of investment capital. Most postwar Texans were poverty-stricken, credit and capital for any purpose were scarce, and banks, few and weak in every part of the state, were practically nonexistent in East Texas. Capital to develop the industry would have to come from the North and East. Although the Piney Woods had a number of experienced lumbermen, many of whom had been engaged in the business for a decade or more, few if any were familiar with major lumber manufacturing and distribution methods. Local operators would develop the needed expertise in time, but in the mid-1870s none of them apparently thought in terms of big mills and quality lumber.[22]

A more obvious problem was transportation. The war and Reconstruction had halted railroad construction, and in 1875 the state had only sixteen hundred miles of mainline track. No rail line ran west of Fort Worth, and railroad transportation had not yet reached San Antonio. More important for prospective lumber entrepreneurs, existing railroads merely skirted the edges of the great pine forests. The heart of the Piney Woods had not yet been penetrated. The coastal towns could ship finished lumber by boat along the Gulf Coast to southwest Texas, Mexico, and elsewhere. The Texas and New Orleans Railroad and in Northeast Texas the Gould lines (the Texas and Pacific and the International and Great Northern) provided rail shipment to a few fortunate communities in Central Texas, but most of the new, rapidly growing prairie communities had to depend on wagons for the long, slow, costly haul from the East Texas mills. Traveler Frank Taylor described the situation near the end of the decade: "Before the construction of the railroads one of the greatest difficulties the prairie settler had to encounter

was the scarcity and extreme cost of lumber. It sold as high as $60.00 to $70.00 per thousand feet and was often hauled hundreds of miles by ox teams."[23]

Although the delivery of finished lumber to the consumer was the principal transportation problem that faced the Texas lumber entrepreneur in the 1870s, not unrelated was the problem of hauling logs to the mill in sufficient quantity to maintain a regular and economically profitable schedule. If the logger depended on mules and oxen for power, the distances must be kept short and the mill close to the source of timber. In practice this meant that the mills remained small and were forced to move frequently. One could float or raft logs down the East Texas rivers, but there were long periods of low water during the year when that was not possible, and logging by river was much less satisfactory in Texas than in the Great Lakes states. A network of tramroads leading to the mills would be much more efficient.[24]

Finally, there was the problem of gaining acceptance of the product. Although builders had used yellow pine for all kinds of construction in the South for more than a century, builders in the East and Middle West and their descendants on the central plains regarded yellow pine as inferior. They objected to its high resin content and thought that it warped excessively and split too easily. As long as the buyer held these views, southern yellow pine could not command the prices that white pine enjoyed in the major markets of the United States. It was only when the great white-pine forests of the East and the Great Lakes states began showing signs of depletion that builders turned to southern yellow pine and began to appreciate its qualities of great flexibility and strength. This change in opinion did not come

[22] Richardson, *Texas: The Lone Star State*, p. 339; Robert S. Maxwell, "Lumbermen of the East Texas Frontier," *Journal of Forest History*, April, 1965, pp. 12–13.

[23] Frank H. Taylor, "Through Texas," *Harper's New Monthly Magazine* 59 (October, 1879):707.
[24] Horn, *This Fascinating Lumber Business*, pp. 29–31; Robert S. Maxwell, "The Pines of Texas: A Study in Lumbering and Public Policy," *East Texas Historical Journal* 11, no. 2 (Fall, 1964):77–78.

overnight, but it was apparent during the last three decades of the nineteenth century. The old prejudices had become too expensive.[25]

These conditions and opportunities attracted many experienced lumbermen with investment capital to the great pine forests of East Texas and neighboring Louisiana. Among the first, if not the first, were Henry J. Lutcher and G. Bedell Moore, of Williamsport, Pennsylvania. Lutcher, the son of an immigrant German butcher, was born in Williamsport in 1836 and engaged in the lumbering business from the early 1850s. Moore, the grandson of an early Episcopal bishop of Virginia, was born the same year in New Jersey and attended Dickinson Seminary before moving west and joining Lutcher in a lumbering partnership in 1862. By 1876 they were operating a successful moderate-sized lumber business, but timber in central and western Pennsylvania was being rapidly depleted, and the partners agreed to seek a new location in another state. According to family and company tradition, they considered Michigan or Wisconsin but decided that it was too late to begin operations there in competition with the many established firms. They had heard of the great yellow-pine forests of East Texas and Louisiana, and although sharing the typical white-pine lumberman's prejudice against yellow pine because of its high resin content, they decided to make a trip south to inspect the region and its opportunities. If their findings proved promising, they planned to purchase timberlands and a millsite and transfer their chief operations to the Gulf Coast.[26] They made this

journey in January and February, 1877, and recorded their travels, train connections, hotels, conversations, and reactions in a pocket-sized diary, which has been preserved. In addition to an account of the travelers' activities the diary provides a valuable addition to the picture of Texas in the post–Civil War generation.[27]

Leaving Williamsport on January 11, 1877, the pair journeyed by train by way of Pittsburgh, Columbus, and Indianapolis, reaching Saint Louis early on the morning of January 13. There they stopped at the Laclede Hotel for two days and met several officials of the St. Louis, Iron Mountain and Southern Railroad (later part of Jay Gould's Missouri Pacific system), who gave them information about timberlands and letters of introduction to individuals in Texas who could assist them in their search. On January 16 they started south, stopping briefly in Arkansas and northeastern Texas to inspect timberlands and visit representative sawmills. At Palestine, Texas, they stopped overnight at another Laclede Hotel and talked with various persons said to be knowledgeable about Texas timber.[28] Significantly, as they traveled Moore recorded the condition of the various rivers they crossed and estimated the water levels, banks, stream-

[25] Horn, *This Fascinating Lumber Business*, p. 102; James E. Fickle, *The New South and the New Competition*, pp. 36–39.

[26] Lutcher and Moore Lumber Company Papers, Forest History Collections, SFA Library; interview with Robert P. Turpin, secretary, retired, Lutcher and Moore Lumber Company, August 13, 1963, Oral History Collections, Forest History Collections, SFA Library. Unless otherwise noted, interviews and conversations cited are in these collections, and all interviews were conducted by principal author Maxwell. See also Boyd, "Fifty Years in the Southern Pine Industry"; Hamilton Pratt Easton, "A History of the Texas Lumbering Industry" (Ph.D. diss., University of Texas, Austin, 1947), pp. 123–25; *Burke's Texas Almanac, 1881*, p. 156.

[27] The original diary of Henry J. Lutcher and G. Bedell Moore's trip to Texas in 1877 is in the possession of Carolyn Negley, of San Antonio, a descendant of G. Bedell Moore. Mrs. Negley presented a bound facsimile copy of the diary to author Maxwell, which is now with the Lutcher and Moore Lumber Company Papers in the Forest History Collections, SFA Library. To judge from the handwriting of the diary and other samples of the partner's correspondence, Moore was the keeper of the diary. The author is also indebted to George K. Stephenson, U.S. Forest Service, retired, who read the diary and the manuscript of this chapter and offered a number of helpful suggestions.

[28] Lutcher and Moore, Diary, pp. 1–8, Lutcher and Moore Lumber Company Papers. The St. Louis, Iron Mountain and Southern Railroad, the Texas and Pacific Railroad, and the International and Great Northern Railroad were part of Jay Gould's empire, with headquarters in Saint Louis. The name Laclède (for Pierre Laclède Liguest, founder of Saint Louis) was popular among Saint Louis businessmen. The hotels by that name may have been part of a chain along the route that became the Missouri Pacific. See Edwin C. McReynolds, *Missouri: A History of the Crossroads State*, p. 20; Julius Grodinsky, *Jay Gould: His Business Career, 1867–1892*, pp. 353–54.

Henry J. Lutcher, pioneer lumberman, Lutcher and Moore Lumber Company, Orange. *Authors' collections.*

G. Bedell Moore, pioneer lumberman, Lutcher and Moore Lumber Company, Orange. *Authors' collections.*

flows, and drifts. They had used the rivers in Pennsylvania for transporting logs, and they obviously hoped to use the streams to drive logs economically at their new location. Indeed, Moore appraised every river they crossed for its potential for rafting logs.[29]

At Houston, which they reached on January 20, the travelers stopped at the Hutchins Hotel, which Moore described as *the* hotel in the city, a very comfortable and commodious house that provided them with a room for three dollars a day. There they spent the weekend, walking about the city and perhaps attending church services. They were interested in Buffalo Bayou, which they described as narrow but deep at the head of tide and navigable by fair-sized steamboats from Galveston. Houston, they learned, was the center of the railway net, ten railroads having terminals in

the city. Moore estimated the population at twenty to twenty-five thousand.[30]

During the trip south Lutcher and Moore had apparently stopped only twice at hotels, the Laclede Hotel in Saint Louis and at the small hotel of the same name in Palestine. Evidently they planned their travel so that they slept on the coaches and spent their days inspecting mills and timber and talking to local citizens. Since both men were over forty years old, they must have been in excellent physical condition, for they gave no hint that they considered the trip unduly tiring. The various timber stands and mills that they had inspected up to this time had not impressed them in comparison to their own operations in Pennsylvania. Those had been merely preliminary

[29] Lutcher and Moore, Diary, pp. 1–14, Lutcher and Moore Lumber Company Papers.

[30] Ibid., p. 14. Apparently Moore obtained his estimates of population from local boosters, for in many cases they were somewhat high. The 1880 census listed the population of Houston at 16,513. See *Texas Almanac and State Industrial Guide, 1978–1979*, pp. 189–93.

examinations, however. Now the partners prepared to inspect the great virgin longleaf-pine stands along the Neches and Sabine rivers. Their findings would determine the success of their mission.

On Monday, January 22, Lutcher and Moore left Houston for Beaumont on the Texas and New Orleans Railroad. After an eight-hour trip of eighty-three miles they arrived at 4:00 P.M. and spent the remaining hours of the day inspecting mills and meeting individuals who might assist them in their mission. In Houston they had made the acquaintance of Judge J. F. Crosby, vice-president of the T&NO, who gave them letters of introduction to several prominent citizens of Beaumont. These gentlemen, according to Moore, "greeted them cordially" and helped them arrange for a horseback trip through the longleaf forests.[31] While they were in Beaumont, they became interested in the cypress-shingle mills and the methods used to float cypress timber down the Neches to the mills. Apparently they encountered pecky cypress for the first time and were intrigued by its appearance and characteristics. On inquiry, the mill owners quoted prices for cypress shingles at $2.25 to $3.00 a thousand.[32]

The next morning they departed for a close inspection of the pine forests. They rowed up the Neches about three miles to Beard's Bayou, where they transferred to horses. Accompanied by a guide, young James Ingalls, Jr., they rode north, crossed Pine Island Bayou on a ferry, and continued north for about eight miles. They spent the night with William

Hooks, who introduced the Pennsylvanians to East Texas country fare: bacon, sweet potatoes, corn bread, and coffee. Moore was not impressed and described the food as "of poor quality." All the next day they rode northward through pine forests relieved by occasional breaks of farmland. The lumbermen could not have failed to be impressed by the magnificent stands of longleaf pine, many trees towering 150 feet and 5 feet in diameter. Other travelers at this period remarked on the "parklike" forest floor, which was virtually free of underbrush and scrub hardwood.[33] Most of the acreage that they traversed was old-growth virgin longleaf pine that had never known an ax or a saw. As the T&NO officials at Beaumont had told them, timberlands were cheap, and stumpage was almost a drug on the market since it did not sell well. Transportation was lacking except for the Neches River. Mill owners in Beaumont were floating logs downriver from distances as great as fifty to one hundred miles.[34]

After a long day's travel of almost thirty miles the men stopped to spend the night with Isham Sheffield. There, in addition to bacon, corn bread, and coffee, they feasted on "cold turnips, tops and all mashed up together." Although Sheffield warned the travelers that he had only frugal fare, Moore described the meals as "miserable" and complained that they were overcharged at $1.25 each, "more than we were charged anywhere that we stayed in the woods."[35]

[31] The Texas and New Orleans Railroad was built from Houston to the Louisiana state line at Orange before the Civil War. At the time of Lutcher and Moore's visit Orange was still the eastern terminal. The road was not completed to New Orleans until 1881. See St. Clair G. Reed, *A History of the Texas Railroads and of Transportation Conditions under Spain and Mexico and the Republic and State*, pp. 84–87, 229.

[32] Lutcher and Moore, Diary, pp. 15–17, Lutcher and Moore Lumber Company Papers. Pecky cypress had a textured appearance that Moore thought made it look as though it had been "eaten lengthwise of the tree by insects." This apparently did not affect its strength or durability.

[33] Lutcher and Moore, Diary, pp. 18–21, Lutcher and Moore Lumber Company Papers. No further mention was made of Ingalls, but apparently he was an efficient guide, for the party did not get lost. Two years later Frank H. Taylor traveled through the same region and described the East Texas forests as most impressive. See Taylor, "Through Texas," pp. 706ff.

[34] Lutcher and Moore, Diary, pp. 18–19, Lutcher and Moore Lumber Company Papers. Moore was keenly interested in the methods of rafting logs downstream and the water level of the Neches at different times of the year.

[35] The quality of the food was no doubt poor, but the basic fare—corn bread, bacon, and coffee—was standard for rural East Texas residents. The diet had not changed much since the 1850s. See Frederick Law Olmsted, *A Journey Through Texas*, pp. 67–68.

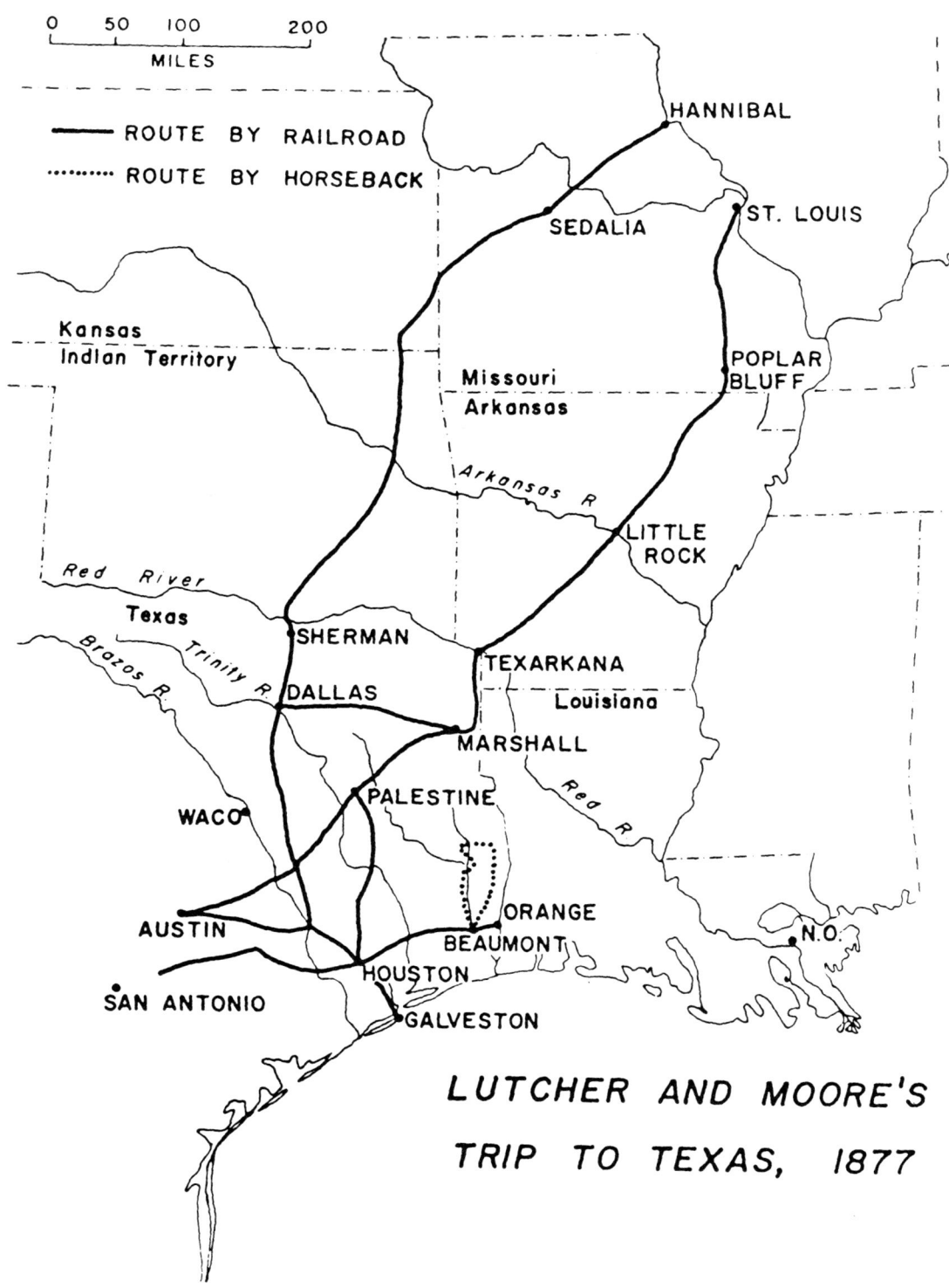

Lutcher and Moore's trip to Texas in 1877, showing the railroad net at that time. Dotted lines trace

On the third day they rode north to Town Bluff, on the Neches, where they crossed the river. They continued north, stopping to have lunch with a Frank Smith, praising his fare as "the best corn bread and sweet potatoes since we had been in Texas." That night they spent with a man named Hamilton, who lived a mile east of Bevilport. There Moore reported excellent accommodations and the "first white bread since leaving Beaumont." Apparently both Hamilton and his wife were intelligent and well read, and the travelers enjoyed a conversation about the merits of various regions of Texas.[36]

On the morning of January 26, Lutcher and Moore with their guide crossed the Angelina River at Bevilport, where they saw the steamboat *Laura* docked and loading cotton.[37] They rode about ten miles in a northerly direction between the rivers, observing and measuring timber, which was of about the same quality and size as the growth they had passed through; here, however, some had been cut and hauled to the river. They turned east, recrossed the Angelina at Morris's Ferry, and rode to Magee's Mill, where they stopped for lunch. Continuing on to Jasper, the partners met a Colonel Dorn, whose name had apparently been given to them in Beaumont. Moore described him as a long-term resident who was well informed about the region. They discussed the longleaf stands they had seen, and Dorn informed them that the timber did not vary much along the same parallel of latitude from the Neches east across the Sabine and into western Louisiana. He told them that the timber on the high ground was larger, with less sapwood than that in the river bottoms.[38]

After an hour or so with Colonel Dorn, they rode another two miles and spent the night with Jehu Bevil. The son of John Bevil, the original white settler of the region, Jehu was an elderly eccentric and "unreconstructed" rebel who enjoyed shocking strangers and newcomers. Moore and Lutcher were properly amazed by their host and his behavior, as Moore's entry for the evening indicates:

Arriving there at dusk, the proprietor [Bevil] invited us in and giving the horses in care of the Negro boy we alighted. The house was built of rough boards and the floor was sanded, the owner was walking around in his stocking feet. On being questioned as to why the floor was strewed with sand, he replied, "The ground is clayey in this neighborhood and I have sand sprinkled on the floor so that the nigger can sweep it easier." Mr. Bevil is 70 years old and is said to have killed several men during the past 20 years. He said he owned 23 Negroes before the war; two of them he had been offered $2,000 apiece for. He did not care about the Emancipation but thought the Government should have paid something for the slaves, even if it was only a moderate sum. He said he had never whipped a nigger in his life, that he himself had never worked any and never intended to. A search was made for the candles to light up the house with, but the key of the trunk in which they were kept was missing. Mr. B said he "kept the candles locked up to keep the niggers from burning them." While we were eating tea a large black woman held a pine torch high in the air for us to see by. On one end of the small room was a large window and at the other end a large door, both wide open, cooling us and the victuals off at the same time; the household did not seem to mind it. Before retiring we desired to know if we could have an early breakfast, and were told we could have it by "sun up." We did not have breakfast until 8:30 and were not on our way until 9 o'clock. This is a fair sample of the habits of some of the people.[39]

After this experience Lutcher, Moore, and young Ingalls turned south and rode back to Beaumont. They observed that the timber between the Neches and the Sabine appeared

---

[36] Lutcher and Moore, "Diary," pp. 24–25, Lutcher and Moore Lumber Company Papers.

[37] The *Laura* was a well-known steamboat on the Neches and Sabine waters. One author has described her as 115 feet long with a beam of 32 feet. See William Seale, "River People," *East Texas Historical Journal* 5, no. 1 (March, 1967): 43–50.

[38] Lutcher and Moore, Diary, pp. 26–27, Lutcher and Moore Lumber Company Papers. In 1880, Jasper, the county seat of Jasper County, was a rural village with a population of five hundred.

[39] Ibid., pp. 27–30. On Bevil see Thomas A. Wilson, *Some Early Southeast Texas Families*, ed. Madeleine Martin, p. 51.

to be larger than that west of the Neches, and denser, with fifteen to twenty mature trees an acre. After spending another night in the forest, they reached the river at Wiess Bluff and rode on into Beaumont, returning about 5:30 P.M. on January 28. In six days they had ridden about 175 miles on horseback, mostly through virgin longleaf-pine timber. The partners were enthusiastic about the trip, impressed by the fine timber stands, and, having left Pennsylvania in midwinter, appreciative of the weather, which Moore described as "all the time fine, the days bright, and the nights beautiful moonlight."[40]

The indefatigable travelers were up the next morning at 5:30 to catch a work engine to Orange. This East Texas border city, standing on the Sabine River just north of the head of Sabine Lake, was already known as a lumber center. There they met, perhaps through letters of introduction from Judge Crosby, several T&NO officials, and Judge Wingate. As mentioned earlier, Wingate had spent a lifetime in the lumber business, and his new sawmill at Orange had an estimated capacity of 20,000 to 30,000 board feet of lumber a day.[41] It was probably the most modern mill in the region.

From Wingate and others the partners learned about the advantages of the Orange region for large-scale lumber operations. They learned that the current of the Sabine was strong most of the year and that the only shallow shoals upstream could easily be bypassed by means of a short ditch or canal. The banks were firm, and timber was available much nearer the water on both the Texas and the Louisiana sides than it was along the Neches. The Sabine was influenced by an eighteen-to-twenty-inch tide and would provide ample

space for freighters to dock and to turn in the channel after loading. Orange was closer to blue water than Beaumont, and the water over the bar at the mouth of the Sabine was a foot deeper than the water at the mouth of the Neches. According to Moore, the transportation of timber to the mill and the shipping of finished lumber by water from Orange would be both feasible and economical. Three sawmills and two shingle mills were operating in Orange, but all were relatively modest operations and would not seriously compete with the mill that Lutcher and Moore planned to construct.[42]

The partners found Orange attractive and pleasant. They remarked on the beauty of the flowers blooming in the gardens in January—verbenas, violets, and other varieties—which made a favorable impression on travelers from the snowbound North. Their landlady, Mrs. Curry, praised Orange as a healthy and desirable place to live.[43] Although Moore did not record his conversations with Lutcher about their decision to locate in Texas, they apparently agreed on Orange as the millsite before they left. Thereafter, their inquiries became more specific, regarding machinery and markets, and they were willing to relax a bit and see part of historic Texas before returning to Pennsylvania.

The next morning (Tuesday, January 30) the partners took the T&NO back to Houston, arriving about twelve hours later. Again they took up quarters in the Hutchins Hotel and spent the next day meeting with Houston businessmen and visiting a sawmill at Harrisburg, which Moore thought was the best mill he had seen in Texas. That evening they caught a train to Galveston. They checked in at the Girardin House, which Moore described as "a comfortable hotel where we had a good bed."[44]

[40] Lutcher and Moore, Diary, pp. 31–32, Lutcher and Moore Lumber Company Papers. On several occasions Lutcher and Moore took time to comment on the fine weather that they were encountering. Perhaps it is fortunate for their venture that they missed the periods of cold, fog, and thunderstorms that East Texas sometimes experiences in winter.

[41] Allen, *East Texas Lumber Workers*, pp. 16, 142; Marcus E. Sperry, *The Port of Orange, Texas, U.S.A.*

[42] Lutcher and Moore, Diary, pp. 33–36, Lutcher and Moore Lumber Company Papers.

[43] Ibid., pp. 36–37; Sperry, *The Port of Orange.*

[44] Lutcher and Moore, Diary, pp. 37–39, Lutcher and Moore Lumber Company Papers. Riding the slow, bumpy T&NO back to Houston, the partners no doubt compared the rail transpor-

The travelers were prepared to enjoy Galveston. They strolled about the city, walked on the beach, and stood on the shore of the Gulf of Mexico, which they had often read about but "never before had seen." Throughout the island they saw that the vegetation was much advanced: peas in blossom, beets and radishes growing in abundance, and orange trees with fruit on them. They strolled on the Strand and remarked on the large number of fine buildings, both public and private, adorning the city. They learned that Galveston had not had an outbreak of yellow fever for almost ten years and that the town was orderly, though fights were not uncommon, and "the revolver [was] used frequently."[45]

Although their trip to Galveston was principally to see the Gulf, the travelers did not neglect business inquiries. They talked to the proprietor of the Lee Foundry about engines, boilers, and other machinery, and Lee agreed to compute the cost of the power plant they wanted and mail them an estimate in a few days. Then they went to a flour mill to inspect one of Lee's engines, describing the cylinder and pillow block as "too high" and the bed plate "too straight." The fact that the mill had been shut down so that the workers could cut the slag out of the boilers, described as a weekly occurrence, was not lost on the experienced lumbermen. They also visited a cotton warehouse and examined a large cotton compress. Knowing nothing about cotton, they were impressed by the machinery, and Moore recorded everything the guide told them about the operations of the compress.[46]

Early the next morning they took a train back to Harrisburg and changed to the Galveston, Harrisburg, and San Antonio Railroad for a visit to San Antonio. Relaxed and enjoying the landscape, the travelers commented on the beauty of the prairie sunset. "It cannot be described," wrote Moore; "it must be seen to be appreciated."[47] Whenever the train stopped, the partners got off and visited lumberyards, estimated lumber stocks, and inquired about the prices of various items. Along the entire route they gauged the potential market in the blacklands for quality lumber, produced with big-mill efficiency, from mature longleaf pine. What they found indicated that the market had hardly been tapped. The temporary end of the line was at Marion, though railroad workers expected to have the road completed into San Antonio, about twenty-five miles away, within a month. They continued their journey to San Antonio by stagecoach and arrived eight hours later after a rough, rather unpleasant ride.[48] They stopped at the Menger Hotel in Alamo Plaza, described by Moore as "a very good house."[49]

They spent the next day, Sunday, seeing the historic sights, visiting churches, and attending Episcopal services. In the afternoon they strolled about the city, checked lumberyards, and noted the San Antonio River as it wound its way through the city, crossed by "numerous iron bridges." Even then the river was a central feature of the Bexar capital.[50] When they

[47] Lutcher and Moore, Diary, pp. 45–46, Lutcher and Moore Lumber Company Papers.

[48] Here as elsewhere Lutcher and Moore were keenly interested in the types, prices, and marketing arrangements of lumber sold in Central Texas. The GH&SA did reach San Antonio in February, 1877. Soon afterward Collis P. Huntington, of the Southern Pacific, acquired the road and extended the line to El Paso. Eventually it became a link in the Southern Pacific system, which ran from New Orleans to San Francisco. The entire road became known as the "Sunset Route." See Reed, A History of the Texas Railroads, pp. 190–98.

[49] More than one hundred years later the Menger Hotel is still in operation and continues to be "a very good house."

[50] Moore had evidently read an account of the Texas Revolution before making the trip. He described San Antonio as an old town, "half German and the rest Mexican." Lutcher and Moore, Diary, pp. 52–54, Lutcher and Moore Lumber Company Papers.

tation in East Texas with the more advanced railroads in Pennsylvania. Perhaps they even heard the nickname given to the T&NO by local wits: "Time No Object." See Reed, A History of the Texas Railroads, pp. 225–29.

[45] Despite the yellow-fever threat, Galveston in 1877 was both the chief port and the largest city in the state. According to the 1880 census, Galveston had a population of 22,248 (Moore estimated it at about 30,000). Texas Almanac, 1978–1979, pp. 189–93.

[46] For more on Galveston and the Galveston Wharf Company, which was perhaps the cotton compress they visited, since it was the largest operation in the city, see Reed, A History of the Texas Railroads, pp. 283–85, 490–92.

learned that track laying on the GH&SA had almost reached the city, they resolved to avoid the long, uncomfortable coach ride back to Marion. Thus, on Monday morning they took a hack to the rail site and hitched a ride on the construction train to Marion. The trip was not without excitement, for they had to cross a branch of the Guadalupe River on a temporary bridge that resembled a modest roller coaster. Moore described the experience:

The approach to this "low water bridge" was a steep down grade to the edge of the river, and the track rose to a corresponding height on the opposite side, the impetus gained by going down hill took us up the other side. Such a fearful plunge for an engine and train to make and keep the rails I never saw. We sped like lightning down the hill, shot across the bridge and up the track on the other side, and gave a sign of relief when we were safe again on the level.[51]

At Marion they boarded the afternoon train for Houston and arrived at 6:00 the next morning, ready for a strenuous day. They visited a mill in Harrisburg that was for sale and tried to see A. C. Winch, the attorney in charge of the sale of the mill. He was not in, and they left a message asking him to write to them in Williamsport, giving them the price of the mill and terms. They also visited several Houston foundries and machine shops and asked the owners to send them prices and bids on engines, boilers, and other machinery that they would need. By late afternoon they had completed their rounds and were ready to begin their journey home. They had decided not only to move to the Orange area but also, if possible, to purchase their machinery in Houston to save an expensive freight bill from the Northeast.

They continued to estimate the potential lumber market as they traveled north.[52] Their

return trip reveals much about the state of rail travel in Texas during the 1870s. They boarded the Houston and Texas Central train bound for Dallas and changed at Hempstead to the T&NO bound for Austin. Again they spent the night on the coach, getting what sleep they could and probably being aroused each time the train stopped at a station. On the way they continued gathering statistics about retail lumberyards and business opportunities.[53]

Reaching Austin at 6:00 A.M., the partners refreshed themselves, had breakfast, and set out to see the capital city of Texas. Moore described Austin as "a fine city, good natural drainage, solid stone and brick buildings including a beautiful new court house." He noted that the city was surrounded by hills and was said to be "very healthy," with a population estimated at thirteen thousand. As usual they sought out retail-lumberyard owners, one of whom, Captain Willett, encouraged their venture and estimated that "Sabine" (longleaf) lumber, properly manufactured, would bring five dollars a thousand board feet more on the Austin market than the loblolly then being offered.[54]

Apparently armed with a letter of introduction, the Pennsylvania lumbermen called on Governor Richard B. Hubbard at the state capitol. Hubbard had only recently assumed the governor's office—in December, 1876, when Richard Coke had resigned to accept election to the United States Senate. Hubbard greeted them warmly and welcomed them to Texas. He expressed concern lest unscrupulous "agents" defraud innocent strangers in sales of lands and townsites. Lutcher and Moore were far from innocents, however, and moreover

---

[51] Despite the uncertain schedule and the excitement of the river crossing, the partners had cut in half the time of the return trip to Marion. Apparently anything was preferable to the long stagecoach ride over rough roads. Ibid., pp. 55–56.

[52] *Burke's Texas Almanac, 1880* lists J. C. C. Winch as a Houston attorney (p. 148). This was perhaps the same man. The same edition of the *Almanac* carried an advertisement of the

Lutcher and Moore Lumber Company featuring spearheaded pickets (p. 68). Evidently L&M lost no time putting that popular item into production.

[53] Lutcher and Moore, Diary, pp. 57–60, Lutcher and Moore Lumber Company Papers. It is probable that neither the H&TC nor the T&NO carried sleeping cars for these runs at the time.

[54] Ibid., pp. 60–61. Moore found Austin an attractive, growing city that would provide an expanding market for quality lumber. With a mill at Orange Lutcher and Moore would be able to ship directly to Austin on the T&NO.

were not interested in speculative purchases.[55]

Before noon of the same day they again boarded the train, this time the I&GN (now the Missouri Pacific), for the trip home. They changed at Hearne, catching the H&TC for Dallas. Again sleeping on the coach, they reached Dallas at 7:00 the next morning and set about making desirable contacts in the shortest possible time. Local lumber dealers estimated that about 15 million board feet of lumber were sold annually in Dallas, mostly from mills on the Texas and Pacific Railroad, which ran west from Shreveport and Marshall. Moore described Dallas as a city of about twelve thousand and "a great many cheap buildings."[56]

Perhaps of greater significance was the call that the partners paid on William Cameron. The forty-three-year-old Cameron was an established owner of a number of retail lumber yards in Missouri, Kansas, and Indian Territory, as well as Texas. In addition to Dallas, Cameron also had yards in Fort Worth and Waco. Impatient with any "losing proposition," Cameron was recognized as an expert on market trends and lumber futures. He was enthusiastic about the future growth of Dallas and Central Texas. He agreed that the current demand was mostly for cheap lumber but predicted that the market for quality materials would improve as the town matured. As Cameron talked, Moore made notes of his estimates of current production capacity and potential consumption. According to Cameron, there was an annual lumber shortage of 35 to 50 million board feet of lumber beyond the output of existing mills. Additional lumber was shipped from Chicago and the Great

Lakes states. The bright future painted by Cameron was not lost on Lutcher and Moore. They saw an opportunity to be in at the beginning of a bonanza.[57]

They journeyed on the H&TC by way of Sherman and then to Denison, where they changed to the Missouri, Kansas and Texas Railroad (Katy), which crossed Indian Territory, Kansas, and Missouri. At Hannibal, Missouri, they changed to the Chicago, Burlington and Quincy bound for Chicago. Moore mentioned for the first time on their trip that they took a sleeping car, indicating that sleeping cars were not available during the Texas portion of their trip. They reached Chicago on Sunday, February 10, and attended church services at Grace Episcopal Church. They were much impressed by the many "fine, new" buildings and especially the Palmer House, which they conceded was the "handsomest" they had ever seen. From Chicago they traveled east, encountering snow and freezing rain and very cold winds blowing from Lake Michigan. After a stop in Erie, Pennsylvania, to discuss sawmill machinery with the officials of an ironworks there, they continued through "intense cold," finally reaching Williamsport on February 13, after an absence of nearly five weeks.[58]

[57] William Cameron had a career that rivaled that of his more famous Scottish countryman Andrew Carnegie. Beginning as a poor young immigrant, he acquired a small retail lumberyard on the Missouri, Kansas and Texas Railroad. He expanded his holdings until he eventually owned more than sixty yards; thousands of acres of pine timber, sawmills, and factories for finished wood products; flour and woolen mills; and other properties. He was a great benefactor to the city of Waco, where he lived for many years before his death in 1899. See Webb, ed., *The Handbook of Texas*, 1:275–76; R. J. Tolson, *A History of William Cameron and Co., Inc.*, pp. 33–37; *Waco Tribune Herald*, April 26, 1964; Lutcher and Moore, Diary, pp. 65–67, Lutcher and Moore Lumber Company Papers.

[58] The Missouri, Kansas and Texas Railroad, which was built through Indian Territory with federal assistance, reached Denison, Texas, in 1872. An agreement with the H&TC created virtually a through line from Houston to the Middle West with convenient connections to Saint Louis and Chicago. At that time it was one of only two rail lines linking Texas with the North and East. See V. V. Masterson, *The Katy Railroad and the Last Frontier*, pp. 164–92; Lutcher and Moore, Diary, pp. 73–75, Lutcher and Moore Lumber Company Papers.

[55] Ibid., p. 62; Richardson, *Texas: The Lone Star State*, p. 228.

[56] At Bremond the H&TC had a branch line running to Waco. Moore was interested in this rapidly growing town and obtained information about its economic potential. It is unlikely because of time and train schedules that the travelers were able to visit Waco. In contrast to their view of Austin, Lutcher and Moore were not favorably impressed by Dallas, though they recognized its potential for growth. Lutcher and Moore, Diary, pp. 63–65, Lutcher and Moore Lumber Company Papers.

Once having determined to make the move, Lutcher and Moore wasted no time transferring their principal operations to the South. Before the end of the year they had returned to Orange, purchased a small sawmill, and begun construction of a large mill that was identified by the initials L&M for the next fifty years. Early employees described it as a mill with double circular saws and a gang saw, with clippers, edgers, trimmers, and a steam-carriage operation. It had a capacity of 80,000 to 100,000 board feet a day and was the first truly large, modern sawmill in the state of Texas. Later the circular saws were replaced with high-speed band saws, and a second mill was built on an adjoining site to complement the first.[59]

Lutcher and Moore purchased timberlands on both sides of the Sabine, some as far away as northern Newton County. They bought thousands of acres in Louisiana from the U.S. General Land Office, which was in charge of sales from the Public Domain. They deliberately chose pine stands, avoiding where possible the hardwoods in the low-lying bottoms. They also acquired large cypress stands, particularly along the lower Sabine and in Louisiana. For these magnificent forests of virgin timber the partners paid very modest sums, usually less than three dollars an acre and, for stumpage, as little as fifty cents an acre. Farmers were eager to sell their timberland, and a local doctor informed Lutcher that the land they were buying was a liability, not an asset, because it was so heavily timbered that it could never be cultivated.[60]

In their logging operations the partners used both river and railroad. They cut timber from their own lands along the Sabine and also bought logs from independent operators.

These loggers, many of them part-time farmers, felled trees upstream, dragged them to the riverbank, made them into rafts, and floated them downstream for sale to L&M. At Orange the company constructed a boom that stretched across the river and corralled the logs, channeling them into side ponds and lagoons for temporary storage. Early in their operations Lutcher and Moore built about thirty-five miles of the Gulf, Sabine and Red River Railroad from their Lousiana pinelands to Niblett's Bluff, on the Sabine. Later they built the shorter Mississippi and Pontchartrain Railroad and the sixty-one-mile Orange and Northwestern, which extended to the company's holdings in Newton County. By thus using all available means of transportation, they assured themselves of a steady supply of logs for the mill. More than other Gulf Coast entrepreneurs Lutcher and Moore made use of the knowledge they had gained in their northern lumber operations and adapted it to the conditions of the South.[61]

Lutcher moved at once to Orange and became active in local and civic affairs. His two daughters married young Texans William H. Stark and Dr. Edgar W. Brown, who also became active in the Lutcher and Moore Company.[62] The Lutcher-Stark-Brown family became a dynasty that has been important in Texas and Louisiana affairs for three generations. Moore also moved to Orange but for a time commuted between Texas and Williamsport, where the partners continued their business until 1888. After that time Moore devoted his entire attention to the company's interests on the Gulf Coast. Soon after the turn of the century Moore sold his share of the various Lutcher and Moore companies to the Lutcher

[59] Lutcher and Moore Lumber Company Papers; interview with B. C. McDonough, general manager, retired, Lutcher and Moore Lumber Company, August 13, 1963; interview with Robert P. Turpin, secretary, retired, Lutcher and Moore Lumber Company, August 13, 1963.

[60] Lutcher and Moore Company Papers; Boyd, "Fifty Years in the Southern Pine Industry," pp. 59–67; *Texas Almanac, 1881*, p. 156.

[61] Sperry, *The Port of Orange*; interviews with B. C. McDonough, Robert P. Turpin, and James H. McNamara, land supervisor, retired, Lutcher and Moore Lumber Company, August 13, 1963; *American Lumberman*, August 26, 1916; interview with M. S. Morris, Sabine County, August 5, 1958.

[62] Webb, ed., *The Handbook of Texas*, 2:94, 229; *Orange Leader*, May 29, 1936. Miriam, the older daughter, married William H. Stark. Her sister, Carrie, married Dr. Edgar W. Brown.

The original Lutcher and Moore "upper" mill, opened in 1877, closed in 1930 because of the De- pression. *Authors' collections.*

family and moved to San Antonio. There he engaged in various business enterprises including banking. His heirs still live in the San Antonio region.[63]

Knowing that the domestic lumber market was subject to violent fluctuations, Lutcher and Moore balanced their United States sales with an export trade. They cultivated markets with Mexico, the West Indies, and South American countries and in time developed a considerable market with countries bordering the Gulf of Mexico and the Caribbean. In the early days cargoes of dressed lumber and squared timbers were loaded on barges, taken downriver across Sabine Lake, and reloaded on ships for the longer voyage. Later, after Lutcher had led the fight to have the Sabine dredged and Orange became a deepwater port,

L&M acquired a large fleet of lumber-hauling vessels. It became common to see several lumber freighters tied at the dock near the L&M mill at the same time. Sawed and dressed lumber with the trademark "Lutcher-Orange" became known the world over.[64]

Lutcher and Moore developed a vast complex of enterprises. Soon after the turn of the century in addition to the parent Lutcher and Moore Lumber Company there were the L&M Cypress Company, the Red Cypress Door and Sash Company, the Orange Mercantile Company, and the Orange Ice, Light, and Water Company. Landholdings in the names of the founders and heirs totaled several hundred thousand acres, a large rice plantation, and the L&M Turpentine Company. Other enterprises included the railroads; the export line, including a number of lumber schooners and freigh-

[63] Lutcher and Moore Lumber Company Papers; Webb, ed., *The Handbook of Texas*, 2:229, 659; ibid., 3:922; *Who's Who in the South and Southwest, 1973–1974*, p. 89, Frances J. Hearn to author Maxwell, July, 1977.

[64] *American Lumberman*, August 8, 1903, p. 35; August 14, 1915, p. 58; September 23, 1916; Sperry, *The Port of Orange.*

ters; and a pioneer experiment in kraft-paper manufacture, the Yellow Pine Paper Mill Company.[65]

The city of Orange benefited from the advent of Lutcher and Moore. Members of the family endowed the city with a church, a hospital, a park, a library, and various other benefactions. Most recently the Brown heirs have given the handsome Conference Center to Lamar University. Perhaps more important, L&M never tried to reduce Orange to "company-town" status. At a time when most of the lumber companies in Texas or Louisiana used merchandise checks and profited from forced trade at the company commissary, Lutcher and Moore, according to one early official, made it a policy to pay its employees in cash every Saturday night.[66]

[65] *American Lumberman*, February 7, 1903, p. 38; March 14, 1903, p. 57; November 6, 1915, p. 64; Sperry, *The Port of Orange*.

[66] *American Lumberman*, November 6, 1915, p. 64; Webb, ed., *The Handbook of Texas*, 2:94, 659; ibid, 3:922; interview with Robert P. Turpin, Orange, August 13, 1963; Lutcher and Moore Lumber Company Papers, box 165; *Lutcher Memorial Church Building* (pamphlet, Orange, 1978).

Thus from the five-week visit of Henry J. Lutcher and G. Bedell Moore to Texas in early 1877 developed the first large mill and the first comprehensive lumbering operation in the state. It marked the beginnings of the bonanza lumber era for Texas, which produced more than 59 billion board feet of finished lumber in the next fifty years and ended with the onset of the Great Depression in the 1930s. When Lutcher and Moore came to Texas, such lumbermen as Alexander Gilmer, David Wingate, John Martin Thompson, and W. T. Carter were merely local manufacturers on a modest scale. The reputations of such tycoons as Joseph H. Kurth, Thomas L. L. Temple, and John Henry Kirby were still in the future. William Cameron was primarily a retail lumber dealer. It is remarkable that these two middle-aged Pennsylvanians had the vision to see the potential for large-scale production and marketing of quality lumber in Texas and the boldness to seize the opportunity that the Piney Woods offered in the post-Reconstruction decade.

CHAPTER 3

Whistle in the Piney Woods

Listen to the lonesome wail of the locomotive in the Piney Woods.—Robert S. Maxwell

ESSENTIAL to the development of a large-scale lumber industry is the availability of adequate transportation. The older American lumber centers in New England, the northeast, and the Great Lakes states depended on rivers. As the Texas industry grew, it became evident that the rivers could not provide reliable transportation for the greater part of the East Texas pine region. If Texas lumbermen were to rival their counterparts in Louisiana and Arkansas or compete with the rapidly growing lumber industry on the West Coast, they needed a railroad network that would make all sections of the Piney Woods accessible to logging operations. The construction and consolidation of this rail net made possible the lumber bonanza in Texas during the half century after 1875. Its conception and growth provide an interesting and important aspect of the saga of the Lone Star State.

Most of the early sawmills in Texas were along the rivers near the Gulf Coast. Their owners had access to water transportation both to bring logs to the mill and to ship the manufactured lumber by barge or ship to such cities as Galveston, Corpus Christi, and Brownsville. Inland sawmill operators such as Peter Ellis Bean and Frost Thorn, of Nacogdoches, and the Thompson family, of northern Rusk County, faced long, expensive, difficult hauls by wagon and oxen if they tried to sell their lumber products beyond their local area. Only mills on or near navigable streams enjoyed the low-cost water transportation that justified shipping lumber any considerable distance from the Piney Woods.[1]

To many lumbermen, especially those who had had experience in the Great Lakes states, location near a free-flowing river was a requirement for logging and lumbering. Before them was the example of such early lumber leaders as Frederick Weyerhaeuser and Isaac Stephenson, who used northern rivers such as the Chippewa and the Menominee to drive and store their logs and then shipped the lumber products downriver to urban markets. This pattern of cutting the timber along the upper reaches of the stream, collecting the logs at the riverbank, and then driving or rafting them downstream to the mill was recognized as standard practice by the veteran lumberman who moved south after the Civil War. The northern rivers were predictable. They carried a reliable volume of water at all times, they froze over in the winter, and with the spring thaw they rose and with their swift currents successfully carried the logs downstream to the mills. Thus the Great Lakes states lumberman had a great free conveyor for his log supply, which he exploited (and fought with other lumbermen to control) until the timber supply was gone. Important lumber-manufacturing towns grew up at the mouths of the rivers, such

[1] Bernice Lockhart, "Navigating Texas Rivers, 1821–1900" (master's thesis, St. Mary's University, San Antonio, 1949).

as Marinette, La Crosse, and Oshkosh, Wisconsin, and Saginaw, Michigan, and even Rock Island, Illinois, and Davenport, Iowa, down the Mississippi.[2]

At first glance it would appear that East Texas rivers, such as the Sabine, the Neches, and the Trinity, could be used in the same way in the Gulf South. On the map the river systems were impressive and extended throughout most of the Piney Woods. They flowed generally south and emptied into the Gulf, where convenient manufacturing establishments and centers for transshipment could be set up. The East Texas rivers, however, were highly unpredictable. They did not, of course, freeze over in the winter, but loggers frequently bogged down during the rainy months of winter and spring. The spring rise could not be certain for the river might flood two or three times or perhaps not at all. On occasion water from spring rains came down the rivers in such a torrent that the logs could not be controlled, and thousands of them were lost. In the summer during a dry year the rivers had almost no current and were virtually useless to drive or raft logs downstream to the mill. The East Texas rivers were also given to obstructions. Sand and silt, logs and tree limbs, and all sorts of debris made navigation difficult and hazardous. So bad did river obstructions eventually become that transportation by steamboat, which had flourished before the Civil War, declined and all but ceased.[3]

True, Lutcher and Moore regularly used the lower Sabine from Niblett's Bluff to raft their log supply to Orange. Mills at Beaumont and Houston and in the Galveston Bay area also used the lower reaches of the rivers with some degree of success. Even here, however, the rivers were unreliable, and in the interior of the Piney Woods (north of the second tier of counties from the Gulf) the great stands of longleaf, loblolly, and shortleaf pine stood virtually untouched. Would-be entrepreneurs awaited a more reliable system of transportation. The coming of the railroads to the Piney Woods helped solve their problems. Rail transportation both public and private became an essential part of the developing lumber industry in Texas during the bonanza period. In turn the East Texas forests promised lucrative freight traffic to the enterprising builder.

As late as 1875 Texas railroads did little more than skirt the edges of the great East Texas forests. The pre–Civil War roadbed of the Texas and New Orleans had been rebuilt from Houston to Beaumont and Orange but was not yet connected with the Louisiana metropolis. A branch of the International and Great Northern, soon to become a property of Jay Gould, ran north from Houston to Palestine and thence to Longview, where it joined another Gould line, the Texas and Pacific. This transcontinental road had one line that extended from Shreveport to Dallas and another that ran from Texarkana to Sherman, with a connecting line from Marshall to Texarkana. And that was all. Most of the heavily forested region of East Texas was still inaccessible by rail, and transportation depended on wagons drawn by oxen, mules, or horses over often impassable roads. By 1875 the need for an adequate rail net that would open the resources of the region for the benefit of all was apparent.[4]

The first railroad to penetrate the heart of the Piney Woods was the Houston, East and West Texas Railway. This romantic indepen-

[2] For interesting accounts of logging and lumbering in the Great Lakes states, see Ralph W. Hidy, Frank E. Hill, and Allan Nevins, *Timber and Men: The Weyerhaeuser Story*; Robert F. Fries, *Empire in Pine: The Story of Lumbering In Wisconsin, 1830–1900*.

[3] See Cleo F. Evans, "Transportation in Early Texas" (master's thesis, St. Mary's University, San Antonio, 1940); Lockhart, "Navigating Texas Rivers"; Robert E. Mills, "Navigation of the Trinity River" (master's thesis, Sam Houston State Teachers College, Huntsville, 1943); *Texas Almanac, 1860*; *Texas Almanac, 1911*, p. 126.

[4] Frank H. Taylor, "Through Texas," *Harper's New Monthly Magazine* 59 (1879):706; St. Claire G. Reed, *A History of Texas Railroads and of Transportation Conditions under Spain and Mexico and the Republic and State*, pp. 745–46. From Shreveport the Gould lines had an outlet to New Orleans on the Texas and Pacific, and Gould owned the Iron Mountain and Southern Railway, which connected Texarkana with Little Rock and Saint Louis.

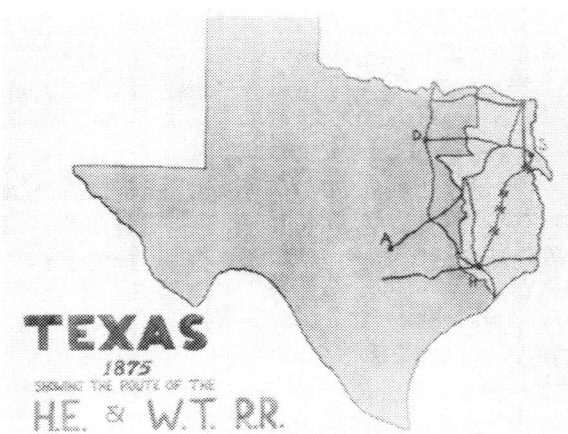

Route of the Houston, East and West Texas, 1875.
Courtesy of Stephen F. Austin State University,
Forest History Collections.

dent narrow-gauge road with the expansive
name ran northeast from Houston to Shreveport, meeting the various lines in both cities.
Its builder and chief financier was Paul Bremond, a well-known Houston businessman
who had earlier built the Houston and Texas
Central Railroad. Bremond was an interesting
individual. Born in New York City, he became
a hatter, but after two unsuccessful ventures he
moved to Galveston in 1839 and engaged in a
general-merchandise business there. Within a
few years he moved to Houston and became
involved in various enterprises, both alone and
in association with such well-known Housto-
nians as William Marsh Rice, T. W. House,
and William A. Van Alstyne. Although the
construction of the H&TC was successful, the
road was heavily in debt, and Bremond lost
control of the company in 1858. He continued, however, to be financially involved in this
and other railroads radiating from Houston.
In addition he acquired important parcels of
Houston real estate, which rapidly appreciated
in value as the town grew into a city.[5]

    [5] Conversation with Edith Bremond Koch, Houston, Oc-
tober 30, 1958; Benajah Harvey Carroll, *Standard History of
Houston, Texas, from a Study of the Original Sources,* pp.
239–40.

Bremond was a prominent pillar of Houston
society, married three times, and reared a large
family. His son and several of his sons-in-law
became associated with him in his various
business enterprises, especially the HE&WT.
Bremond was also an earnest believer in spir-
itualism. Under his leadership a spiritualist society met regularly in Houston and on occasion entertained visiting practitioners from the
East. He attended spiritualist conferences in
New York and believed that his personal guide
and spirit contact was Moseley Baker, a veteran of the Texas Revolution who had fought
at the Battle of San Jacinto. According to Bremond, it was Baker who advised him to build
the HE&WT. When several years went by
without action, the spirit of Baker prodded
him to proceed with the work. In 1875, Bremond was sixty-five years old, a plump, bright-
eyed, old-fashioned gentleman who wore a full
beard and affected clothes with an English cut.
In any company he was an arresting figure and
gave the impression that he could command
millions. On his presence, his personal credit,
and very little cash Bremond built the first railroad through the Piney Woods.[6]

Receiving a charter for the HE&WT in
1875, Bremond began construction the next
year. The route stretched north and east of
Houston, across the San Jacinto, and on to
Cleveland, Livingston, and Moscow. Because
Bremond believed that a narrow-gauge road
would be quicker and cheaper to build and
would operate as efficiently as a standard-
gauge road, he ordered the HE&WT to be
constructed with a three-foot gauge. It was
true that this reduced construction time, but it
also prevented interchange of rolling stock
with a standard-gauge road at either Houston
or Shreveport. For a railroad dependent on a
heavy, low-value commodity such as lumber
for most of its freight tonnage, this was a
serious disadvantage. By the time the line
reached Cleveland, five sawmills were operating along the right-of-way with a combined ca-

    [6] Robert S. Maxwell, *Whistle in the Piney Woods: Paul Bre-
mond and the Houston, East and West Texas Railway,* pp. 4–8.

Whistle at Cotton Square. Lufkin was a thriving railroad town. *From a painting by Raymond Ryan;* *courtesy of Lufkin Printing Company.*

pacity of 100,000 board feet of lumber a day. The lumber was shipped south by rail and sold in Houston.[7]

The pine forests provided not only the chief traffic for Bremond's road but the sinews for the line itself. Pine and oak were used for ties; longer timbers were made into trestles to span the Trinity, Neches, and Angelina rivers in turn; and the pert little puffer billy engines burned wood as fuel.[8] As the HE&WT built northward, new mills sprang up and provided more tonnage for the road. Bremond's road also hauled manufactured goods, machinery, foodstuffs, and consumer items north to the villages and towns along its way. The HE&WT was the lifeline of the Piney Woods.

Financial problems slowed his progress, but Bremond continued to build additional track and put it into operation. By the spring of 1882 the road reached central Angelina County,

and Bremond laid out and established Lufkin, named for his chief construction engineer. Within another year he had spanned the Angelina River and entered Nacogdoches, the Spanish town dating from the early eighteenth century. There, where sawmilling had been a profitable operation since the days of Peter Ellis Bean, several steam-driven circular mills promptly began advertising pine lumber at lowest prices, to be delivered to the buyer on the cars of the HE&WT.[9]

The rest of the line was completed south from Shreveport and northeast from Nacogdoches, the rails meeting at the Sabine River, where a new iron bridge was constructed for the crossing. Bremond died in May, 1885, but the long-awaited first train finally ran from Houston to Shreveport early in 1886. With the opening of the through route the HE&WT provided connections with the Texas and Pa-

[7] *Houston Telegraph*, April 18, 1878; November 16, 1878; Maxwell, *Whistle in the Piney Woods*, pp. 10–12.

[8] Bremond purchased the *Girard* and the *Centennial*, two gaily painted locomotives used at the Philadelphia Centennial Exposition, as the initial rolling stock for the HE&WT. Maxwell, *Whistle in the Piney Woods*, p. 11.

[9] Ibid., pp. 19–21; conversation with J. Frank Summers, 1955, Oral History Collections, Forest History Collections, Stephen F. Austin State University Library, Nacogdoches; hereafter cited as SFA Library. Unless otherwise noted, interviews and conversations cited are in these collections, and all interviews were conducted by principal author Maxwell.

cific and other lines that could carry the traveler or loaded freight car to New Orleans, Saint Louis, and other points north, east, and south. Bremond's road was the first to open the resources of the Piney Woods to the outside world. Its completion marked the beginning of a new era.[10]

Long before its tracks reached the Sabine, the HE&WT had acquired a nickname that stayed with it for more than thirty years. Because of its short, bobbing narrow-gauge cars, its up- and downhill roadbed, its "out-of-breath" engines, its tendency to jump the track, and, above all, its proclivity for "stopping behind every stump," the road was promptly labeled the "Rabbit." And the "Rabbit" it remained for more than a generation. It is ironic that the farmers and businessmen along the route, the very people most eager for Bremond to build his railroad through their counties and towns, were the first to denounce the line's shortcomings and deride the narrow gauge and its antiquated equipment. If the train was woefully late, the word would be passed around among the crowd at the depot that "the Rabbit has jumped the track again." If the train pulled in on time, some wag would convulse the "boys" with the observation that it was "yesterday's train." On the route of the HE&WT were four small towns that gave rise to a popular song, "Teneha, Timpson, Bobo, and Blair."[11]

In an era in which all railroads had nicknames and the initials in their names were given as bizarre a translation as possible, it was perhaps inevitable that the HE&WT acquired a tagline that became famous throughout the country. The HE&WT, it was said, was the perfect example of the traveling salesman's nightmare—"Hell Either Way Taken."[12] Per-

haps it was. With the sway from its narrow-gauge base and the pitch built into its short tracks and rough roadbed, plus the hand brakes and manual couplings that prevailed until the turn of the century, the drummer (traveling salesman) must have looked upon any extended trip on Bremond's road with grim foreboding.

The counterpart of the HE&WT was the Texas and New Orleans Railroad running from Beaumont to Dallas and crossing Bremond's road at Nacogdoches. The T&NO, owned by the Southern Pacific Company, built the line through the heart of the Piney Woods between 1882 and 1903, buying existing companies along the route and building new trackage as necessary. Among the short lines absorbed in this merger were the Sabine and East Texas Railway and the Texas Trunk Railroad. The T&NO ran northwest from Beaumont through Kountze, Woodville, Rockland, Zavalla, Dunagen, Nacogdoches, Jacksonville, Athens, and so on, to Kaufman and Dallas. Like other roads of the period, the T&NO acquired nicknames. Some referred to it as "Time No Object" or "Turnips and New Onions." To some it was the "Wooden Axle Road." Regardless of the name the local wags coined to describe it, the T&NO opened up a great area of East Texas pinelands not otherwise accessible, and a number of major lumber companies established mills along its route. It also opened a direct route to the rapidly growing region of Dallas and North Texas for the lumber products of the Piney Woods.[13]

A third major railroad that penetrated the Piney Woods was the Atchison, Topeka and Santa Fe. The interest of the Santa Fe in the opportunities of the East Texas forests began casually and almost by accident. In the 1890s, John Henry Kirby was beginning to acquire timberland and sawmills that were eventually to make him the largest producer of yellow pine in the South. To connect his properties, he

---

[10] The Sabine River bridge was officially completed on January 28, 1886, and the first through trains ran the next day. See Maxwell, *Whistle in the Piney Woods*, pp. 28–31.

[11] *Dallas Morning News*, March 13, 1962; *Lufkin News*, July 15, 1962.

[12] B. A. Botkin and Alvin F. Harlow, eds., *A Treasury of Railroad Folklore*, p. 506.

[13] Reed, *A History of Texas Railroads*, pp. 230–33; Maxwell, *Whistle in the Piney Woods*, pp. 62–63.

had chartered and built about sixty-two miles of railroad between Beaumont and Rogansville called the Gulf, Beaumont and Kansas City Railway. Kirby, needing more capital to develop his holdings, borrowed money from the Santa Fe through eastern bankers. As one railroad historian described the relationship, the ensuing transaction was only natural:

It was just before the turn of the century that the Santa Fe and Kirby came together. He needed more money for his lumber business, but did not need this railroad. The Santa Fe had the money, needed the lumber business and knew how to build and operate railroads, so it bought the G.B.&K.C. from him and also agreed to lend Kirby $2,000,000. . . . The agreement was most advantageous to both.[14]

About the same time the Santa Fe began to acquire and build east–west trackage to connect the GB&KC with its main line at Sommerville. By 1902, using several independent short lines as a nucleus, the Santa Fe had completed the branch, which ran from Sommerville to Navasota, Conroe, Cleveland, Kountze, and Silsbee. The road provided an outlet for Texas longleaf lumber, especially Kirby products, on Santa Fe tracks to the plains states, the West Coast, and even the Chicago market. Although indirect in point of mileage, much of the lumber from the eastern tier of Texas counties reached market by this route.[15]

The Santa Fe, under the name Gulf, Beaumont and Great Northern, extended this branch line northward through the magnificent longleaf pine forests to Jasper and San Augustine in 1902 and to Center a year later. The company then purchased a much-reorganized short-line road running south from Longview and Carthage called the Texas and Gulf. Although the T&G had a line extending as far south as Grigsby (on the western edge of San Augustine County), the Santa Fe officials elected to build new track from Gary to Center, thus closing the gap. This construction,

completed in 1908, extended the Santa Fe line through the eastern tier of Texas counties from Beaumont north to Longview, where it made connection with the Missouri Pacific and the Texas Pacific. Along this route many independent sawmills and lumber camps were established, but in a very real sense it remained "Kirby country" throughout the era.[16]

With these three main trunk lines serving as a skeleton, short lines soon crisscrossed the region to provide a complete railroad net for the entire East Texas region. Many of the major lumber companies built their own connecting lines, secured charters as common carriers from the Texas legislature, and operated the companies as separate entities. Some enterprising builders and progressive towns promoted the construction of connecting short lines to advance or protect their economic interest. Other major railroads built lines into the Piney Woods to tap the lumber traffic. A special case was the East Texas branch of the Missouri, Kansas and Texas Railway, commonly referred to as the "Orphan Katy." This story takes us back to Jay Gould.

The decades of the 1870s and 1880s found Jay Gould at the peak of his career. Already well known as a market speculator, stock manipulator, and promoter, Gould had established himself in Saint Louis and had become a major railroad owner and builder. He sought to dominate the rail system in Texas and to build a transcontinental line to the Pacific Coast. Locally Gould had achieved notoriety because of his antagonism to Jefferson, Texas. Gould had refused (perhaps wisely in view of the subsequent history of the Red River raft) to make Jefferson his eastern terminal and had even diverted his Marshall–Texarkana branch line to the edge of town. At various times Gould owned or controlled not only the Missouri Pacific, the Texas and Pacific, and the International and Great Northern but also the Wabash, the Union Pacific, the Denver and Rio

[14] Reed, *A History of Texas Railroads*, pp. 293–94.
[15] Maxwell, *Whistle in the Piney Woods*, pp. 51–52.

[16] Ibid., p. 65; Reed, *A History of Texas Railroads*, pp. 293–94.

Network of railroads in East Texas during the bonanza era (1911), showing Kirby Lumber Company connections. *Authors' collections.*

Grande Western, and the Missouri, Kansas and Texas railroads as well. A shrewd judge of legal loopholes in the days before the establishment of the Interstate Commerce Commission, Gould frequently milked one rail line for the benefit of another, usually to the advantage of his personal holdings in the Missouri Pacific and the Texas and Pacific.

In 1881 a local group in Trinity County chartered and began construction of a short line, the Trinity and Sabine Railway, to tap the rich longleaf stands in Trinity and Polk counties and provide a connection between the I&GN and the HE&WT, which was building north from Houston. The T&S group completed thirty-eight miles of road from Trinity to Corrigan, where it made connection with Bremond's road. At once, in December, 1882, Gould bought the short line in the name of the Missouri Pacific and on the same day, in his dual capacity as president of both lines, sold it to the Missouri, Kansas and Texas. The MKT, popularly known as the "Katy," had no use for this thirty-eight-mile strip of road, for it had no main-line track closer than a hundred miles away at Waco. The line was extended farther eastward to connect with Colmesneil on the T&NO, and this track too was sold to the Katy. It is not known what Gould paid the T&S for the road, but, according to the biographer of the Katy, the MKT paid the Missouri Pacific $40,000 a mile. Large and prosperous lumber companies such as Thompson and Tucker and William Cameron were set up on the route, chiefly to the benefit of the I&GN and the HE&WT. No doubt Jay Gould chortled when he relinquished control of the Katy in 1888, for he had effectively looted the Katy and fastened "Orphan Katy" on the mother company for the benefit of his other properties. The Katy was stuck with operating a feeder line for its rivals until it finally disposed of the unwanted orphan after 1920.[17]

The railroad provided the only fast and dependable transportation in and out of the Piney Woods. Here lumberman J. Lewis Thompson boards the Katy at Willard for a business trip. *Courtesy of Stephen F. Austin State University, Forest History Collections, Thompson photo albums.*

Eventually the East Texas branch of the MKT joined another short line based at Trinity. This road had its beginnings in 1905 as the Beaumont and Great Northern, chartered and owned by William Carlisle and Company. Carlisle was an important lumber magnate originally from Wisconsin who owned two large mills with headquarters at Onalaska. According to local tradition, Carlisle had a special fondness for the name Onalaska, for he had previously given the name to mill towns in Wisconsin and Arkansas. Later, after he had cut out his Texas timber, he founded a fourth Onalaska on the West Coast.[18]

The Beaumont and Great Northern ran from Trinity to Livingston, a distance of about thirty-three miles. It passed a number of towns, mostly sawmill sites with such romantic names as Pagoda, Sebastapol, Fitzvann, Luce, Carlisle, Pennell, Onalaska, Kickapoo, Blanchard, Vreeland, and East Tempe. The line planned to build to the Gulf but never extended beyond Livingston, though it built

[17] Vincent V. Masterson, *The Katy Railroad and the Last Frontier*, pp. 227–35, 282–84; Reed, *A History of Texas Railroads*, pp. 382, 483–84; Maxwell, *Whistle in the Piney Woods*, pp. 56–57.

[18] Conversation with W. S. Brame, Livingston, 1963.

Locomotive of the Chronister Lumber Company, Angelina County, ca. 1925. *Courtesy of Stephen F.* *Austin State University, Forest History Collections.*

westward to Weldon. After it changed hands several times, the parent Katy Railroad purchased it in 1912, proposing to link it with its main line at Waco and build eastward to a saltwater terminal at Port Arthur. The plan was never carried out, for the Katy went into receivership. Later, during World War I, the federal government intervened to freeze the properties in the status quo until after the war.[19]

In 1923 the Katy disposed of both the B&GN and the Orphan Katy to Colonel R. C. Duff, who had previously owned the B&GN. He reorganized the lines and secured a new charter in the name of the Waco, Beaumont, Trinity and Sabine Railway. Although he planned to extend the road to the Gulf, he was unable to do so, and in 1930 the line again went into receivership. In the minds of most local residents the initials stood for the rather ignominious nickname "Wobblety-Bobblety, Turnover, and Stop." Nevertheless, the WBT&S opened up an important area of

virgin longleaf pine between Trinity on the I&GN and Livingston and Corrigan on the HE&WT. Along both branches of the "Wobblety-Bobblety" sawmills were so close together that it was said that one was never out of the sound of a mill whistle.[20]

Another independent line that built into the Piney Woods was the Cotton Belt, more officially known as the Saint Louis Southwestern Railway. Like many other Texas railways, this road has an interesting history and has borne a number of names. It began in 1877 as a narrow-gauge local road from Tyler to Big Sandy on the T&P and was first called the Tyler Tap. About the same time another local company under the name Rusk Transportation Company secured a charter to build to Tyler by way of Jacksonville. Within a few years the company's funds were exhausted, and the railway was sold to the Kansas and Gulf Shortline, which completed the road to Tyler, and ex-

[19] Masterson, *The Katy Railroad*, p. 280; Reed, *A History of Texas Railroads*, p. 484.

[20] Conversation with T. L. Epperson, Trinity, June 5, 1962; Reed, *A History of Texas Railroads*, pp. 484–87; Masterson, *The Katy Railroad*, pp. 282–84; Lucius Beebe. *Mixed Train Daily*, pp. 96, 111, 318, 332.

Angelina and Neches River locomotive no. 2, ca. 1918. *Courtesy of Stephen F. Austin State Univer-* *sity, Forest History Collections.*

tended it to Lufkin in 1885 and eventually to White City, in San Augustine County. Like the HE&WT it was for a time a narrow-gauge line that depended largely on lumber and lumber products for its traffic. Along its route were a number of sawmills, including the Angelina County Lumber Company, the Chronister Lumber Company, and the Lufkin Land and Lumber Company.[21]

As large lumber companies moved into the Piney Woods, many owners built logging railroads that carried logs to their mills and connected with the main trunk lines for shipment of finished lumber products. Often these logging roads secured charters as common carriers and operated a general freight and passenger business in addition to their primary functions as tramroads for the parent lumber company. Many of these short lines bore expansive names and acquired even more bizarre

nicknames. As a rule the smaller and more insignificant the road was, the more outlandish was its name. Perhaps the ultimate in railroad names was coined by Thomas E. Durham, who operated a small tramroad in the Marshall area extending about five miles from the T&P tracks at Hallsville to the Durham mill. This line, whose total motive power consisted of one Shay locomotive, bore the name "The Great Sweetgum Yubadam and Hoo Hoo Route."[22]

There were literally dozens of company-owned railroads that were combined tramroads and common carriers. Most of them were built between 1890 and 1920. It would not be feasible to attempt to discuss all the roads, but a few can be described as representative and with rather interesting histories in themselves. That three of the four examples were still operating as of this writing may pro-

[21] Maxwell, *Whistle in the Piney Woods*, pp. 58–60.

[22] Reed, *A History of Texas Railroads*, pp. 503–504.

Mainline locomotive and crew with a string of log cars. *Courtesy of Stephen F. Austin State University, Forest History Collections.*

Shay locomotive and a log train. *Courtesy of Stephen F. Austin State University, Forest History Collections.*

vide a long-range perspective on their history.

The Angelina County Lumber Company built an independent short line in 1900, called the Angelina and Neches River Railroad, which ran generally eastward from the company mill at Keltys, on the Cotton Belt west of Lufkin, crossing the HE&WT and the T&NO before crossing the Angelina River on a pine trestle. From there it passed through a section of magnificent longleaf pine and terminated at Chireno, a small village near the Attoyac River. The A&NR served as a common carrier, operating a "mixed train daily," hauling miscellaneous freight and transporting the few passengers in a coach or caboose. The parent company also used the tracks for log trains to the mill. The connections with two other railroads gave the Angelina County Lumber Company a choice of routes in addition to the Cotton Belt.[23]

Thomas Lewis Latane Temple and the Southern Pine Lumber Company built a short line called the Texas South-Eastern with headquarters at Diboll. Begun in 1900, the road originally ran eastward, opening up the timber stands in the Neches River bottoms. The line was later extended west of Diboll to Blix and Vair and thence to Lufkin, with a spur running into Houston County, ending at Bluff City. For a time the TS-E ran a daily mixed train that served the farmers of the area as well as providing transportation for the logging crews and their families. Among the local citizens the initials TS-E were said to stand for "Tattered, Shattered, and Expired." In addition to service as a common carrier the TS-E also performed as a logging railroad for the parent company and various Temple enterprises.

Perhaps the most famous of the Texas short lines was the Moscow, Camden, and San Augustine Railroad. This lively little "cracker-barrel" road was the property of W. T. Carter and Brother Lumber Company, chartered in

Shay locomotive no. 2, W. T. Carter and Brother Lumber Company. *Authors' collections.*

1898. Like most of the other roads in the Piney Woods, it served as both a common carrier and a logging road for the parent company. The main line actually ran only between Moscow and Camden, a distance of about seven miles, but several spur lines extending almost the length of Polk County were used exclusively for logging. For many years the "mixed train daily" of the MC&SA made the run, powered by a 2-6-0 Mogul locomotive, which had earlier seen service in the Panama Canal Zone. Passengers rode in a bright-red combine with a green roof and rattan seats that had once carried commuters on Long Island. This picturesque little train rumbled through the closely wooded lowlands between the HE&WT junction at Moscow and the W. T. Carter mill town, Camden, on a regular daily schedule. From all over the world rail fans and nostalgic visitors came to ride behind the venerable iron

[23] Records of the Angelina and Neches River Railroad, ACLCo. Papers, Forest History Collections, SFA Library; conversation with W. S. Scott, general manager, A&NR, June 6, 1962.

Frost-Johnson Lumber Company's mainline loco-
motive, Nacogdoches, ca. 1925. *Courtesy of Ste-* *phen F. Austin State University, Forest History* *Collections.*

horse on the shortest mainline track in Texas. In addition to serving as a vital logging road for W. T. Carter and Brother Lumber Company and an outlet for its lumber products, the MC&SA developed into a considerable tourist attraction, discussed wherever railroad enthusiasts gathered.[24]

Another representative shortline railroad was the Nacogdoches and Southeastern, which extended from the old Spanish town east-southeast across the Attoyac to a point on the Santa Fe called Calgary. The road was built by the Hayward Lumber Company and was extended by the Frost-Johnson Lumber Company, which bought the Hayward properties. Begun in 1903, the N&SE became a common carrier in 1905. In addition to running log trains to the mill and extending spur lines deep into the forest to facilitate operations, the N&SE ran a mixed train three days a week in each direction, serving such places as Woden,

Oil Springs, Littles Chapel, Camp Worth, Harmony, and Calgary (N&SE timetable). The train not only hauled lumber and lumber products in and out of the Piney Woods but also carried foodstuffs, fertilizer, garden produce, cotton, cattle, and even automobiles. The shortest route to take from San Augustine to Nacogdoches and points south on the HE&WT was to go to Calgary and change to the N&SE. As the "boys at the depot" would say, its initials stood for "Never Say Early." It very seldom was.

Sawmill hands sang a song about the N&SE that went something like this:

Oh that train—
That Southeastern passenger train!
I'm gonna buy me a ticket
as long as my arm,
I'm gonna ride that train
all night long!

Owned entirely by the parent lumber company and joined by a series of ever-changing spur lines, the N&SE was a profitable money-maker for many years and served the typical

[24] B. A. Botkin and Alvin F. Harlow, eds., *A Treasury of Railroad Folklore*, p. 242; Beebe, *Mixed Train Daily*, p. 111; *Dallas Morning News*, May 20, October 22, 1956.

THE NACOGDOCHES AND SOUTHEASTERN RAILROAD COMPANY

Southward   Northward

No. 1 Mixed Train Leave Tuesday Thursday Saturday	Distance from Nacogdoches	Station No.	Time Table No. 13 Cancels Time Table No. 12 Effective 12:01 a.m. November 18, 1928 STATIONS	Length of Siding	Fuel, Water Y Phone	No. 2 Mixed Train Arrives Tuesday Thursday Saturday
8:00 a.m.	.0	1	Nacogdoches	Yard	P	2:30 p.m.
8:05 a.m.	1	3	Hayward	Yard	YWX-P	2:15 p.m.
8:20 a.m.	7.3	4	Hampton	900	P P	2:00 p.m.
8:35 a.m.	9.9	6	McClures	747	P	1:50 p.m.
8:45 a.m.	11.1	7	Woden	1133	S P	1:40 p.m.
9:05 a.m.	15.3	8	Oil Springs	Y	S P Y	1:25 p.m.
9:25 a.m.	20.6	9	Littles Chapel	1100	P	1:05 p.m.
9:40 a.m.	22.8	10	Pauls Valley	Spur 400	P	12:55 p.m.
9:55 a.m.	26.3	11	Attoyac	1160	P	12:35 p.m.
10:30 a.m.	32.6	12	Camp Worth	Y	Y P S	12:10 p.m.
10:55 a.m.	38.6	13	Harmony		P	11:15 a.m.
11:10 a.m.	42.3	14	Calgary	Y	Y P S	11:30 a.m.
Arrive Tuesday Thursday Saturday						Leave Tuesday Thursday Saturday

SPECIAL RULES

Train No. 1 has right of way over Train No. 2. Trains must not exceed speed limit of six miles per hour within the city limits of Nacogdoches, and over La Nana, Carrisso Creek, and Attoyac River bridges. All trains must come to a stop and flag Chireno Highway crossing just east of Pauls Valley. Switch at east end of Hayward yard must be left set for skidway track. N. & S.E. trains or engines must not use the Santa Fe tracks at Calgary until first ascertaining that all past due trains have arrived and then only under proper protection.

P. Phone, Y. Wye, W. Water, X Fuel, F. Stop on Signal, S. Stop.

Approved:
H.W. Whited, V.P. & G.M.

P.L. Williams, Supt. and Traffic Manager

Timetable for typical shortline railroad, with scheduled passenger service, ca. 1928. *Authors' collections.*

Nacogdoches and Southeastern mainline locomotive and crew, ca. 1927. *Courtesy of Stephen F. Austin State University, Forest History Collections.*

dual purpose of logging road and common carrier. It was abandoned in 1954.[25]

One reason for the multiplicity of shortline railroads in the East Texas Piney Woods was the pattern of "divisions" or allowances in railroad-traffic tariffs that developed during the 1890s. Because of the keen competition, trunkline railroads allowed divisions of the freight charges with the originating short lines in return for shipping lumber and lumber products over their roads instead of those of rivals. This practice provided a means for concealing rebates, for the haul of the so-called originating line was often very short, sometimes no more than a few hundred feet. The same service could as easily have been performed by the trunkline railroad with a spur or siding and a switch engine.[26]

It was to take advantage of this favorable situation that many of the owners of large, per-

manent mills incorporated their tramroads under separate and distinctive names, obtained charters as common carriers from the state of Texas, and performed various common-carrier services such as regular freight service and more or less regular passenger service available to the general public. Wherever practicable, the shortline owners built their roads to connect with two or more trunkline railways and thus provided themselves with alternate routes for shipping their lumber and lumber products to market. Thus the Kurth-owned Angelina and Neches River Railroad could provide connections for Angelina County Lumber Company products with the Cotton Belt, the T&NO, or the HE&WT. The Nacogdoches and Southeastern, owned by the Frost-Johnson Lumber Company, made junctions with the T&NO and the HE&WT and eventually with the Santa Fe at Calgary. The Temple common carrier, the Texas and Southeastern, not only provided the obvious outlet by way of the HE&WT at Diboll but connected with the Cotton Belt at Lufkin and, by using the Groveton, Lufkin and Northern and the Orphan Katy, could ship its products by way of the

[25] Minute Book, Nacogdoches and Southeastern Railroad, 1905–26, Forest History Collections, SFA Library; Reed, *A History of Texas Railroads*, pp. 461–62; *Tyler Morning Telegraph*, January 13, 1949.

[26] Interstate Commerce Commission, Bureau of Statistics, *Interstate Commerce Commission Activities, 1887–1937*, p. 241.

Log trains in the Piney Woods. *Courtesy of Stephen F. Austin State University, Forest History Collections.*

I&GN and the Missouri Pacific system. Those lumber companies without common-carrier shortline railroads under their control, even large, permanent lumber manufacturers, found themselves at a disadvantage in competing for large lumber sales in the major markets. Those with alternate outlets were in an even better position to effect arrangements for the shipment of their products.[27]

By 1906 this practice had become widespread throughout the Southwest. In that year the passage of the Hepburn Act by Congress raised serious doubts concerning the legality of such divisions by the trunk lines with so-called tap lines, and complaints poured into the Interstate Commerce Commission. The commission undertook an extensive investigation

of the problem and in 1912 handed down a sweeping decision directing the trunk lines to "cease and desist" from granting divisions or rebates to tap lines for hauling lumber products for "proprietary companies" but to allow divisions to the short lines for any nonproprietary freight transported.[28]

The decision by the ICC threatened the lumber companies and their subsidiary railroads with serious loss of revenue. The allowances paid usually ranged from $.015 to $.05 per 100 pounds out of the trunk line's earnings under the group-lumber rate. It was estimated that the aggregate amount of allowances paid throughout the country was no less than $50 million a year. As might be expected, many companies appealed the decree, first to the

[27] See Texas railroad map no. 2 in Reed, *A History of Texas Railroads,* pp. 499–504.

[28] See "tap-line case," *United States* v. *Louisiana and Pacific Railroad Company et al.,* 234 U.S. 1: 1185–98 (1913).

United States Commerce Court and eventually to the United States Supreme Court.[29]

In 1913 the Supreme Court set aside the original ICC order but affirmed the authority of the ICC to remedy any abuses that had grown up. "Where the divisions of joint rates are such as to amount to rebates or discriminations in favor of the owners of the tap lines," said the Court, "it is within the province of the Commission to reduce the amount so that a tap line shall receive just compensation only for what it actually does."[30]

In accordance with this decision the commission vacated its original ruling and issued new orders defining the maximum divisions that could be allowed to tap lines. In subsequent litigation the commission's directives were upheld. This series of findings, appeals, and court decisions had the long-range effect of clarifying the status of tapline railroads, especially those owned by an industrial complex such as a major lumber company. Recognized as common carriers with the right to participate in divisions of through traffic originating with the tap line, the shortline railroads were, nevertheless, firmly under the supervision of the ICC. The divisions and allowances received by the short lines were much more modest than those of the years before 1910.[31]

The consolidation of rail lines and technological improvements greatly increased the railroads' capacity for service to the lumber industry in East Texas. By 1900 virtually all the independent roads and most of the tap lines had converted to standard gauge. Such inventions as automatic couplers, air brakes, and block signals improved transportation and promoted safety. The Southern Pacific acquired the HE&WT, which, with the T&NO (and in 1932 the Cotton Belt) made the SP the dominant system in the Piney Woods. The same road led the way in the changeover from wood to coal and then to oil as fuel for its locomotives.[32]

Unlike the lumber industry in the Great Lakes states, the industry in the Piney Woods of Texas and western Louisiana developed parallel with the development of the railroads. Logs made their way from the forest to the mill on the spur lines and the main line of the company's logging road. The finished lumber products went to market on the tap lines and the main trunk lines of the Southern Pacific, Santa Fe, and Missouri Pacific. Machinery, supplies, and personnel came into the lumber towns and camps the same way. The lonely whistle of the locomotive in the Piney Woods heralded the growth of the commercial lumber industry in East Texas.

[29] Ibid., pp. 1194–95.

[30] Ibid., p. 1196.

[31] O'Keef v. United States, 240 U.S. (1915); 294 U.S. Bureau of Statistics, Interstate Commerce Commission, *Interstate Commerce Commission Activities, 1887–1937*, p. 242.

[32] Maxwell, *Whistle in the Piney Woods*, pp. 42–50.

CHAPTER 4

In the Woods

They cut down the old
pine tree,
and they hauled it
away to the mill.—folk song

LOGGING operations in their simplest and most primitive form would appear to be as elementary and obvious as indicated by the lines above, from the old folk song. Yet a pine log twenty feet long with an average diameter of eighteen inches weighs one ton, and an efficient double-band mill requires about nine hundred to one thousand logs a day to operate at capacity.[1] With the advent of the big mills in permanent locations the requirement changed from bringing the mill to the timber to bringing the logs to the mill, sometimes dozens of miles. The logistics of moving a thousand tons of large, cumbersome, unwieldy logs such distances from remote, almost inaccessible forests every day were, to say the least, formidable. Until the arrival of the railroads the task was often impossible.

The southern logger had little in common with the lumberjack of American folklore and tradition. The eastern or Great Lakes states logger was pictured as a wild, hard-working, hard-drinking, devil-may-care bachelor. He was colorful and picturesque in his red toboggan cap, his bright-red shirt, and his heavy calked boots. His work was largely seasonal in the northern states, and in the fall he made for the logging camp, where he lived in a bunk-house with other members of his crew. There they consumed mountains of flapjacks, bacon, beans, potatoes, and venison steak and also enjoyed roistering horseplay that could suddenly turn into a bloody free-for-all. During the winter, often in subzero weather, they cut the tall white pines and snaked them to the river. In the spring came the log drive downriver, when the logger demonstrated his expertise with the cant hook, the pike pole and the peavy or competed in birling. Once the logs were safely corraled in a boom and sorted for the respective owners, the logger headed for the nearest large town and celebrated the completion of the year's work with a spree that rivaled the antics in the cow towns and mining centers of the West. The lumberjack's classic boast that "I can run faster, jump higher, squat lower, and spit farther than any jack in camp" was made in good humor but also in good faith. He seriously thought that he could.[2]

The Texas logger was not like that at all. His garb was much less colorful and lacked any special lumberjack distinction. Instead of cap, plaid shirt, waterproof pants, and calked boots, he was likely to wear bib-front overalls, brogans, an old shirt, and a large hat, ei-

[1]Computation of U.S. Forest Service, Southern Forest Experiment Station, Nacogdoches office; computations of Laurence C. Walker and Ellis V. Hunt, Department of Forestry, Stephen F. Austin State University, Nacogdoches.

[2]Stewart H. Holbrook, *Holy Old Mackinaw;* Robert F. Fries, *Empire in Pine: The Story of Lumbering in Wisconsin, 1830–1900,* pp. 24–60. For lumber camp food see Joseph R. Conlin, "Old Boy, Did You Get Enough of Pie?" *Journal of Forest History* 23 (October, 1979): 164.

Flatheads fell an old-growth pine, ca. 1910. *Cour-*
tesy of Stephen F. Austin State University, Forest   *History Collections.*

ther felt or straw. He was a family man. The "front," or logging camp, instead of a row of bunkhouses and a cook hall, was a village. There he lived with his wife and children in a frame house from which he departed for work every morning and to which he returned every night. His diet was somewhat less robust, consisting of beans, molasses, corn bread, coffee, salt pork, and some fresh beef or pork.[3]

In further contrast to his Great Lakes states cousin, the Texas or Louisiana logger worked the year round at his job and relied less heavily on water transportation to deliver the logs to the millsite. Nor did he use an iced skidway, for there was no ice. More likely his primary transportation was a slow, plodding bull or mule team. Although probably no more pious than the northern lumberjack, the southern logger lacked the opportunity for hell raising and weekends of vice. Such glittering centers of sin and entertainment as Saginaw, Michigan, were nonexistent in the South. He did not engage in contests to display his skill with the cant hook, the peavy, or the pike pole, nor did he compete in birling. And he probably never heard of Paul Bunyan.

Before the railroads penetrated the Piney Woods in the 1880s and 1890s, the early lumber-manufacturing centers ranged along the Gulf Coast. The mills depended on the inland forests for their log supply and on the rivers and creeks for transportation. In the early days before 1900, the loggers imitated northern techniques, which they adapted to the southern climate. The Sabine and Neches rivers and their tributaries were the principal arteries, and the mills were clustered on the banks of the two rivers, chiefly at Orange and Beaumont.

On the Sabine were the pioneer firm Lutcher and Moore and also the Alexander Gilmer, the Miller-Link, and the David R. Wingate lumber companies. In and near Beaumont, on the Neches, were such early operators as Simon Wiess, the Texas Tram and Lumber Company, the Reliance Lumber Company, and the Beaumont Lumber Company, the last headed by J. F. Keith. With considerable holdings up the Sabine and the Neches they employed their own lumberjacks and contract loggers to cut and deliver pine logs to their mills.

Many prominent lumbermen got their start "running logs" on these rivers. Around 1880, Peter A. Doucette was engaged in cutting logs on Village Creek and running them, either loose or in rafts, down that stream and the Neches to the mills at Beaumont. The younger Wiess, William, cut and ran logs to his father's mill above Beaumont. William A. Fletcher got his start running logs down the Neches for Beaumont lumber companies.[4]

One such old-time logger and river driver was M. S. ("Pud") Morris, of Sabine County, who as late as 1960 was still hale and hearty though nearing ninety. Morris's lumbering career went back to the 1880s, when as a boy in his early teens, he began assisting in logging operations. There was always a market for fine logs cut from the virgin longleaf pine timber that abounded in Sabine County. In those days, he recalled, the logger cut down the tree with an ax, using a saw only to cut the trunk into desired lengths. The limbs were also hewed away with the ax. After the trees had been felled, the loggers dragged the logs to the Sabine, using oxen or mules, and piled the logs on the riverbank.

When they were ready to go downriver, the loggers would bind the logs into rafts and set out for Orange. Morris recalled one such trip, in which six men took ten or twelve rafts downriver. Each raft was thirty or forty feet long and consisted of ten or twelve logs bound together with chains or ropes or fastened with pegs. On one raft they placed a cook shack and a small boat, which they brought along to enable them to tend the rafts and keep them going. When they arrived in Orange, they sold

---

[3] Ruth A. Allen, *East Texas Lumber Workers: An Economic and Social Picture, 1870–1950*, p. 121.

[4] James Boyd, "Fifty Years in the Southern Pine Industry," *Southern Lumberman* 144 (December 15, 1931): 59–67.

Choppers and buckers clear the fallen trees of branches and cut the trees into logs of the desired length. *Courtesy of Stephen F. Austin State University, Forest History Collections.*

the entire lot to Lutcher and Moore. The party returned to their Sabine homes by steamboat. As Morris recalled, the last such trip that he made with rafts was about 1905 or 1906.

During the same period and a few years later they floated loose logs down the Sabine, first warning the lumber companies to be on the lookout and have their booms in place. On other occasions they felled trees and dragged or hauled the logs to the bank of the Sabine, where they sold them on the spot to agents from the lumber companies at Orange—L&M, Alexander Gilmer, Orange Lumber Company, Sabine Tram Company, or Miller-Link, which then assumed responsibility for moving them. This contract business had largely died out before World War I.[5]

[5] Conversation with M. S. Morris, Sabine County, August 5, 1958.

The river itself helped curtail and finally end this traditional method of log transportation. Above the head of tidewater at Niblett's Bluff on the Sabine and at Wiess Bluff on the Neches, water levels were uncertain and traffic became constantly more hazardous. No doubt the very effort to use the rivers as they were used in the North hastened their obstruction. Soon after the turn of the century steamboats abandoned efforts to serve the ports on the upper reaches of the rivers. Mud, silt, snags, logs, rafts, and other debris increasingly clogged the channel. Unlike the Penobscot, the Saginaw, and the Chippewa rivers of the North, the rivers of East Texas and western Louisiana were ill-suited for driving logs.

"Pud" Morris's story was a fairly typical one. Only the finest pines near the river could be profitably cut and sent downriver. By the turn of the century the railroads had invaded even such remote forest regions as Sabine County, and the big mills on the Gulf Coast turned to rail transportation for most of their log supply. Perhaps more important, big new mills were constructed nearer at hand.[6]

Most of the inland lumber companies followed a similar pattern in organizing their lumber operations at the turn of the century. A woods crew was composed of forty to sixty men under the direction of a timberman and under the direct supervision of a woods foreman. The crews were subdivided into fallers and buckers, skidders, loaders, tram crew members, and the steel gang. Living in the company town or in the camp at the company front, most of the crew members rode the empty log train to the site of operations, usually arriving by daylight or shortly afterward. They worked a long day—ten to twelve hours —and usually did not return home until after dark.

The timber cutters went by a variety of names. They were fallers (or fellers) and buckers, flatheads, swampers, or choppers. They set the pace, for all the rest of the work was dependent upon them. Choppers usually worked in pairs and more often were paid on a piece-work basis (a rate per thousand board feet they cut) rather than on a set daily wage. The woods boss (sometimes known as "bull of the woods") designated the trees to be cut, if the cutting was on a selective-cutting basis. If the area was to be clear-cut, he simply marked off an area for a team of choppers.[7]

Needham B. Weatherford, veteran logging superintendent for W. T. Carter and Brother Lumber Company, vividly described a typical pair of company flatheads as they came over the hill into the area assigned to them by the woods boss. One member of the team, slightly ahead of the other, carried a two-man crosscut saw, a double-bitted ax, and a whiskey or other bottle filled with kerosene and stuffed with a rag or cork. He was the leader. His companion carried an ax and a measuring stick. The leader would spot a likely tree, lay down the saw, and squint skyward to determine the extent and regularity of the crown, the list of the trunk, if any, and the probable direction of its fall. All experienced loggers prided themselves on their ability to bring down a tree in a desired space, where it would clear other mature timber and spare seedlings and young trees. Indeed, many expert cutters could literally drive a stake into the ground with the falling trunk of the tree.[8]

Once having made his mathematical calculations on the best and safest direction to drop the tree, the lead flathead would wield his ax to chop out the wedge or undercut on the side he wanted the tree to fall. Most fallers were proud of the smoothness and accuracy of their undercut and would wager with any visitor on their skill. This notch might bite into the tree as much as 20 to 30 percent, depending on the size of the tree. This done, they set to work on the opposite side with the crosscut saw. This was back-breaking work and a constant quarrel went on between the woods boss and choppers over the height of the stump to be left. A sort of rule-of-thumb evolved by which no stump would be higher than its diameter where it was cut. As the sawing proceeded, the men drove wedges into the cut behind the saw to keep it from binding. When the trunk was nearly severed, the tree usually gave a slight shiver (which was sufficient signal to stand clear) and then crashed over in the desired direction.[9]

[6] An exception to this trend was Lutcher and Moore, which continued to log by both rail and river until the onset of the Depression in 1930.

[7] Nelson Courtlandt Brown, *Logging: Principles and Practices in the United States and Canada*, p. 102. In a later period timber spotters or markers preceded the choppers and marked with paint all trees that were to be cut.

[8] Conversations with Needham B. Weatherford, logging superintendent, W. T. Carter and Brother Lumber Company, July 25, 1958, October 13, 1959; Oral History Collections, Forest History Collections, Stephen F. Austin State University Library, Nacogdoches; hereafter cited as SFA Library. Unless otherwise noted, interviews and conversations cited are in these collections, and all interviews were conducted by principal author Maxwell.

[9] Conversation with Needham B. Weatherford; conversation with James La Rue ("Rue") Wilson, timber-crew supervisor, Angelina County Lumber Company, July 28–29, 1958.

A logging crew with mules, wagon, and carts poses during the day's work. *Courtesy of Stephen F. Austin State University, Forest History Collections.*

Once the tree was on the ground, the choppers proceeded to convert the trunk into logs. While one man was chopping off the limbs and larger knots from the main stem, the other was measuring the trunk and deciding on the most desirable lengths into which logs should be cut. This task also called for skill and experience, and the manner in which it was done had much to do with the profit of the operation. A number of factors helped determine the most advantageous log length: the mill's need for timbers of a certain length, the methods of transportation available, and the size, conformation, and quality of the tree. In Texas and Louisiana most trunks were cut into either sixteen- or twenty-foot lengths, plus whatever short length was left. When there was a choice, the skill and knowledge of the chief faller was to be drawn upon. For example, it was gener-ally preferable to obtain three sixteen-foot logs rather than two twenties and an eight. It was standard practice to allow a few inches on each log for later trimming. The tops and branches were abandoned to rot on the ground or, at a later period, to be gathered by a clean-up crew.[10]

Once the job of felling and bucking (sawing the tree into logs and removing limbs) was accomplished, the logs had to be moved to track-side or directly to the mill. If the former, the loggers referred to the task as "skidding," "yarding," "cold-decking," or simply "collecting." For this task they almost universally used horses, mules, or oxen—sometimes all of them

[10] Ibid.; Brown, *Logging*, pp. 118–20; Stanley F. Horn, *This Fascinating Lumber Business*, pp. 124–26. Often a single woods crew had as many as a dozen "saws" (pairs of flatheads), who were the aristocrats of the loggers.

A bull puncher with a team of oxen in virgin pine stand. *Courtesy of Stephen F. Austin State Univer-* *sity, Forest History Collections.*

in the same operation. An average-size log (such as the 18-inch-by-20-foot log mentioned at the beginning of this chapter) to be moved on dry, sandy soil could be dragged to trackside by a pair of mules. Larger logs, or those in low, boggy ground required a bull team, often used in tandem. Although powerful, oxen had disadvantages: they were slow, surly, difficult to manage, and more expensive than mules. For these reasons bull teams, very common at the turn of the century, had largely disappeared by 1930.[11] For somewhat longer hauls

loggers used big-wheel, slip-tongue carts or four- and eight-wheel wagons.

Before the development of the power loader the task of loading logs on a wagon or log car was exacting, difficult, and dangerous. The usual method was the "crosshaul," by which a chain, cable, or rope attached to the wagon was placed around the log and then passed over the wagon to a horse or mule team on the other side pulling at right angles to the load. Two men on the ground steadied and guided the log up a pair of poles placed on an inclined plane to the "top loader," the man on top of the wagon, who eased the log into position on the wagon. As the logs piled high in pyramid formation, the job of positioning them became more and more tricky, and more than one top loader was swept from his perch by a sudden swing of the log or had a foot crushed when the

[11] W. T. Carter and Brother Lumber Company, Camden, maintained and used a bull team consisting of five yoke of oxen until the early 1960s, largely for sentimental and publicity purposes. In addition to the other difficulties was the problem of finding and training a skilled bull driver. Only one aging employee could expertly handle the bulls, and there was no apprentice. Conversation with Thomas L. Carter, president, W. T. Carter and Brother Lumber Company, July 23–25, 1958.

"Big Wheels." Workmen using a slip-tongue high-wheeled cart to move longleaf logs to trackside.

Courtesy of Stephen F. Austin State University, Forest History Collections.

log rolled over without warning. The man with the team, of course, had to coordinate his orders with the signals of the top loader. By this method three men plus the team driver could load 10,000 to 20,000 board feet a day. The chief advantages of the method were that all of the equipment needed was at hand and that it was simple.[12]

A revolution took place with the introduction of the steam loader just before 1900. The favorite steam loader used in Texas was apparently the McGiffert loader, a stoutly built machine that operated from the railroad on its own trucks and was self-propelled for short distances. Its working parts consisted of a boom and cable, which could be raised, lowered, reeled in or out, or swung. At first men on the ground looped the cables around each end of the log and then signaled the man at the controls, locally called the "loader man," to "take it away." Later an enterprising engineer at the Lufkin (Texas) Foundry invented and patented the Martin grip hook, which greatly simplified loading techniques. With the hook the ground men in the loading crew could merely hook the cable into the ends of the log and let the loader man hoist away.[13] It was said that one Mc-

[12] Conversation with Needham B. Weatherford; Brown, *Logging*, p. 247; Horn, *This Fascinating Lumber Business*, p. 128; conversation with Charles Welch, Brookland, August 5, 1958; conversation between Clyde Thompson and Jesse Parker, Diboll, August 11, 1954.

[13] Conversation with "Rue" Wilson, July 28–29, 1958. Conversation with A. E. Cudlipp, Lufkin Foundry Company, Lufkin, March 8, 1960. As author Maxwell discovered, even the handling of the Martin grip hook required considerable practice and dexterity.

Cross-loading operation on a log car, showing top loader. Man in foreground holds a cant hook.

Courtesy of Stephen F. Austin State University, Forest History Collections.

Giffert loader could load more than 100,000 board feet of logs a day.[14]

The bonanza period in southwestern lumbering began with the arrival of the railroads after 1880. With the building of the Houston, East and West Texas, the Texas and New Orleans, the Santa Fe, and other main lines into the Piney Woods, major lumber operators such as the Temples, the Carters, the Kurths, the Frosts, and the Thompsons followed close behind. As discussed in chapter 3, most of these lumber leaders included in their holdings their own railroads, often expanded to common-carrier status. These lines carried ambitious and often exotic names and printed schedules, but their chief function was to serve as logging roads for the parent mill. Examples were the

Texas South-Eastern (the Temple railroad); the Moscow, Camden, and San Augustine (Carter); the Angelina and Neches River (Kurth); the Nacogdoches and Southeastern (Frost); and the Lufkin, Hemphill and Gulf (Knox) railroads.[15]

Just when the first tramroad was built is uncertain, but tramroads were in operation before the Civil War, some using wooden rails and powered by mules or horses.[16] By the turn of the century, however, many logging railroads were in operation, and many companies

[14] Brown, *Logging*, pp. 249–53.

[15] Robert S. Maxwell, *Whistle in the Piney Woods: Paul Bremond and the Houston East and West Texas Railway*, pp. 51–56; St. Clair G. Reed, *A History of Texas Railroads*, pp. 499–504.

[16] See Reed's discussion of early tramroads in *A History of Texas Railroads*, pp. 499–504. One such animal-powered tramroad operator aptly called his road the TM&C ("Two Mules and Car").

Logging crew rests while log train begins trip from
the forest to the mill. *Courtesy of Stephen F. Austin State University, Forest History Collections,
Thompson photo albums.*

carried on their logging activities with them.
The Frost-Johnson Lumber Company, for example, conducted a representative lumber operation in central East Texas. The company
logged principally by railroads, spur lines,
skidders, and steam loaders and may be considered typical of the large inland lumber
companies, including the Kurths, Temples,
Knoxes, Pickerings, and Joyces. Frost-
Johnson's mainline track stretched south and
east from Nacogdoches across the Attoyac to a
junction with the Santa Fe at Calgary. A
branch line also opened up the dense timber
stands at the confluence of the Angelina and
the Attoyac rivers. When engaged in logging a
region, the company built spur lines at right
angles to the main line every half mile or

so into its holdings. Fallers and buckers operated as described above, but the crew did only
a very limited amount of hauling by mules or
horses. The principal work was done by
skidders.[17]

The invention of the steam skidder is usually
credited to Horace Butters, a white-pine lumberman in Michigan. It was brought south in
the early 1890s. The skidder was a steam engine set on an iron frame and mounted on
tracks for operation on the railroad. It was
equipped with booms, power-driven drums,
and two, three, or four steel cables each up to
eight hundred feet long. The simplest skidder
required a man and a mule to reel the cable out
to the logging operation or to the point where
the logs had been collected. There a man attached the cable to the log or logs with tongs or
a loop (often called "choker"), and at a signal
the skidder man rolled in the cable, dragging,
jerking, and hurtling the log to trackside. A
more sophisticated design eliminated the mule,
providing instead a "rehaul" cable, which by
means of a pulley and spar (a block on a stump
would serve as well) returned the tongs and cable to the logging area. The most spectacular
adaptation of the skidder was the high-lead
system. This involved trimming a spar tree and
rigging a high line on it, together with pulleys
and a haul-back line. This method had the advantage of swinging the logs over any obstruction, but it was ill-adapted to terrain that was
still partly forested. It was seldom used in
Texas, though it was used some in Louisiana
and other southern states. In most Texas lumbering operations the high-lead system lacked
the necessary flexibility.[18]

Next to the railroads, the rehaul skidder and
the steam loader were the two most popular logging machines used by the larger lumber operators in the Southwest. By strategi-
cally placing the spur lines, a company could
quickly log entire sections with a minimum of

[17]Conversations with Clyde J. Woodward, Sr., Frost-
Johnson Lumber Company, 1957–63.
[18]Brown, *Logging*, pp. 209–27.

Steam skidder at work. *Courtesy of Stephen F. Austin State University, Forest History Collections.*

manual or animal work. Mules or oxen were needed only to drag the logs within the reach of cables, and since timber more than three or four hundred yards from the spur line would be logged from the next spur, these minor yarding operations seldom extended beyond a few hundred feet.[19]

There were, however, several objections to the use of skidders. The skidder drew the fire of conservationists and professional foresters because it often destroyed young trees, especially pine seedlings, making reforestation much more difficult. The logging crews disliked the skidder because they were the victims of a speed-up when the skidder was working at top speed. Frequently a four-cable skidder reeled in logs faster than the loggers could work, and the men were constantly admonished by the bosses to hurry. The skidder was also responsible for many accidents in the woods. If the cable came loose, it could lash wildly about, possibly decapitating a worker who chanced to be in the way. The logs were dangerous as they hurtled along the ground or bounced in the air. The man placing the tongs or loop on the next batch of logs occasionally got a hand, a leg, or his clothing caught in the cable if the skidder man started the haul-in prematurely. In spite of the employment of a whistle punk (a boy who gave a signal to the skidder operator that all was clear), this kind of accident was common. Workers, especially black loggers, often hinted that a "mean skidder man" could injure someone and make it look like an accident. Nevertheless, the skidder was not abandoned until selective cutting, preservation of young timber, and reforestation became the rule rather than the exception in Texas lumbering.[20]

In any rail and power-skidder logging operation, the steel gang was constantly laying down rails and ties for a spur track and taking them up again when that section was logged out. In most lumbering companies the steel gang was a peculiar and in many ways independent group. In one large lumber company, for example, the twelve-man steel gang considered that eight rails picked up and laid down on a new spur by each man constituted a day's work. With good teamwork and hard effort the gang could often complete that quota by three or four o'clock in the afternoon. Although usually the gang members could not go home until the first log train went in, they considered their short workday a great advantage, and there was no shortage of men willing to transfer to the steel gang.

The steel gang usually exhibited more team spirit and camaraderie than did most of the other workers in the woods. Perhaps it was because of their special function, perhaps it was because the work called for special strength in an occupation of strong men, or perhaps it was because they had to pull together to handle the heavy rails and ties. Whatever the reason, the bonds of loyalty and unity were strong. Veteran logging workers recall the chant of the steel gang as they put a new spur through the forest:

Pick 'em up
and lay 'em down
Jint 'em back!
Swing 'em high.[21]

Although steel-gang members acknowledged that their work was the hardest in the camp, none of them willingly changed jobs. They stayed together out of pride in their group or in their display of strength and skill—or just because of the prospect of two or three idle hours in the afternoon. Their motive certainly was

[19] Ibid., pp. 172–74; conversation with Needham B. Weatherford. According to Weatherford, the Carter and Brother Lumber Company never used a rehaul skidder and used the one-way skidder for only a brief time. Then the company went back to mules and oxen.

[20] Conversation with "Rue" Wilson, Angelina County Lumber Company, July 28, 1958; conversation with Needham B. Weatherford, October 13, 1959. Some veteran lumbermen, on the other hand, argued that the skidder did very little more damage to the forests and ground than did the many roads and

turnarounds necessary with animal skidding. Conversation with Clyde J. Woodward, Sr., Frost-Johnson Lumber Company, 1962. See also Industrial Accident Reports, 1925, ACLCo. Papers, Forest History Collections, SFA Library.

[21] Conversation with Clyde J. Woodward, Sr., 1962.

Steel gang laying track for a spur line. *Courtesy of Stephen F. Austin State University, Forest History Collections.*

not the pay. Steel-gang workers received no more than the average woods worker—about two dollars a day—and less than the fallers and buckers. In any case, a member of the steel gang considered himself among the elite of the woods crew, and he would not change places with anyone.[22]

The logging railroads varied greatly in construction, maintenance, and rolling stock. Some operated over well-ballasted, adequately graded roadbeds, while others could hardly be described as ballasted at all. The width ranged from standard gauge (4 feet 8-½ inches) down to three feet. Because the HE&WT was narrow gauge until 1894, many of the logging roads in the region were also narrow gauge for a time. The general rule was no ballasting and only a minimum of grading for the spur lines, which were temporary and were taken up again in a few days or weeks. The spur lines often resembled snakes as they disappeared into the forest, weaving and dipping, swerving to avoid standing timber, easing right or left to miss a stump that otherwise would be directly under a rail. Crooked though they were, the spur lines efficiently served their purpose—to provide a cheap, dependable means of getting out the logs. Logging cars spotted at the skidder location in the morning could be picked up in the afternoon by the locomotive and assembled in a train for the long haul to the mill.[23]

[22] Conversations with Angelina County logging workers, July 28, 29, 1958; conversation with W. T. Carter and Brother Lumber Company logging workers, July 26, 1958.

[23] Brown, *Logging*, pp. 329–45; Reed, *A History of Texas Railroads*, pp. 499–504; conversation with "Rue" Wilson, July 28, 1958.

Little rod and piston logging engine working on a spur line. *Courtesy of Stephen F. Austin State Uni-* *versity, Forest History Collections.*

Logging locomotives also varied greatly from company to company. Most of them were conventional rod-and-piston engines that had been converted from old main liners or designed especially for logging work. In East Texas there were still locomotives at work during World War I that had served Bremond's road when the "Rabbit" (the HE&WT) was a narrow gauge. For standard-gauge logging roads the new owners outfitted such locomotives with standard-gauge wheels, and they served until they wore out or were wrecked. Since most of the loggers used light rails—fifty to seventy pounds on the main line and thirty to forty pounds on the spurs—these short, lightweight puffer billies performed well, and many were still bringing in log trains after two, three, or even four decades in service.[24]

Many lumber companies also acquired one or more geared engines for logging operations. The most popular of these was the Shay. Ephraim Shay, who was responsible for the design of the locomotive , was a white-pine lumberman in Michigan. He took his plans to the Lima (Ohio) Locomotive Works, which in the course of more than sixty years after 1880 turned out about twenty-seven hundred of these little work engines. In addition to the geared wheels, the most obvious characteristic of the Shay was its off-center boiler. To compensate for three sets of cylinders and the crankshaft mounted on the right side, the boiler was set over to the left, giving it an unbalanced appearance. It was small, slow, and

[24]Reed, *A History of Texas Railroads*, pp. 499–501; John T.

Labbe and Vernon Goe, *Railroads in the Woods*, pp. 2–10; *American Lumberman*, September 26, 1908; conversation with Needham B. Weatherford, July 25, 1958.

noisy but strong, durable, and very flexible. As one writer described it, "No one who has ever known it could forget it; the sight of a Shay thrashing its way up a grade, lost in an aura of sound and smoke and steam."[25]

Many of the Shays found their way to Texas and the Gulf. The Thompsons had at least three Shays, the Carters one, the Temples one, and the Frosts one. Rival geared locomotives included the Climax (first produced in 1888) and the Heisler (1894). Knox owned a Climax, as did Miller-Vidor, and the Pickerings (in both Louisiana and Texas) had three.[26] Although small, these little "tea kettles" had good pulling power and worked well on the light railroads. It was a common sight to see a Shay (or another small geared engine) noisily making its way to mill at the head of fifteen to twenty-five carloads of pine logs.

Once the individual cars were collected and the train was formed on the main line, it remained for the logging engineer to give the highball (signal to get up steam and take off) and head for the mill. On the evening run (many companies brought in two trainloads a day) the train included a caboose, or crummy, which provided transportation for the woods foreman and all or part of the logging crew. It required about fifteen to twenty carloads of logs a day to keep a double-band mill operating at capacity, and twenty cars were a good trainload.[27] The woods boss always tried to keep at least two weeks' supply of logs in the log pond or stacked on the deck in case cold or wet weather curtailed operations in the woods. The railroad was by far the most efficient means of carrying logs on a long straight-line haul, especially in forests of large, dense timber. Log trains ran at an average speed of ten to twelve miles per hour and frequently carried logs distances of twenty to twenty-five miles. For the logging crew, living in the mill town, it

represented an extra two hours of "commuting" each way.[28]

Indeed, it was the long hours of commuting time that brought about the establishment of the logging camp at the "front." Unlike the logging camps in the Great Lakes states these camps were miniature, semipermanent mill towns. There loggers lived with their wives and children, and company officials provided schools and churches and a branch commissary. For example, the W. T. Carter and Brother Lumber Company maintained such a camp for a number of years at Camp Ruby, at the southern end of Polk County, connected with headquarters at Camden by about sixteen miles of main-line tramroad. In like fashion Frost maintained Camp Pershing in lower Nacogdoches County and later Camp Worth, on the Attoyac. The Southern Pine Lumber Company, of Diboll, maintained a logging camp at Fastrill, in Cherokee County, for almost twenty years, beginning about 1922. It was composed of permanent houses arranged in streets, with trees lining the walkways. Other lumber companies operated similar camps on a semipermanent basis.[29]

A constant menace to everyone interested in the forests, owners and logging workers alike, was fire. Although southern forest fires, especially those in Louisiana and Texas, seldom made the headlines that accompanied the "big burns" in the Great Lakes states or on the Pacific Coast, it has been estimated that before World War II about 5 percent of the forested area of Texas and Louisiana burned each year. According to the Texas Forest Service, in 1926 there were 1,754 reported fires that burned 103,132 acres; in 1928 there were 3,282 fires that burned 336,000 acres; and in 1932 there were 6,211 fires that burned 527,446 acres. These figures are representative of the fire men-

[25] Labbe and Goe, *Railroads in the Woods*, p. 83; Karmer A. Adams, *Logging Railroads of the West*, pp. 72–73.

[26] *American Lumberman*, September 26, 1908, pp. 110–11; Thomas T. Taber III and Walter Casler, *Climax: An Unusual Steam Locomotive*, pp. 63–96.

[27] Conversation with "Rue" Wilson, July 29, 1958.

[28] Conversation with Needham B. Weatherford, July 25, 1958.

[29] Ibid.; conversation with Clyde J. Woodward, Sr., 1961; interview of John Larson with Clyde Thompson, Southern Pine Lumber Company, 1954. Some companies used small prefabricated houses that they could move on flatcars. These camps were much less permanent and much more primitive than those like Camp Ruby.

Engineer guides log train around curve en route to the mill. *Courtesy of Stephen F. Austin State University, Forest History Collections.*

ace at the end of the bonanza period in Texas lumbering. In some years during the 1920s fires were less destructive; in other years they were worse.[30]

State foresters advanced a number of reasons for the large number of fires and the resulting destruction. In 1917, State Forester J. H. Foster listed 1,086 reported fires as caused by lightning (8 fires), railroads (61), lumbering, including trams (94), burning brush (141), hunters and fishermen (101), incendiary (68), other known causes (59), and unknown causes (554). It is apparent from this breakdown—and this is a typical distribution—that most fires were man-made and that a large percentage of those were deliberate.[31]

In the Piney Woods of East Texas, as in most of the rest of the rural South, the natives regularly burned over the woodlands during the winter in the belief that it would improve the spring grass; get rid of ticks, chiggers, snakes, and undergrowth; and "generally make the air cleaner." Efforts by timber owners to prevent all burning or to stop trespassing rarely were successful and only raised the hostility of the local population, who had looked upon the parklike forest floor of the longleaf stands as a

[30] Texas Forest Service, *East Texas Protection Area Forest Fire Statistics*, Texas A&M College, Texas Forest Service Bulletin (1957); William G. Wahlenberg, *Longleaf Pine: Its Use, Ecology, Regeneration, Protection, Growth, and Management*, p. 142.

[31] J. H. Foster, *Second Annual Report of the State Forester*, Texas A&M College, Department of Forestry Bulletin no. 8 (1916), p. 4; Richard G. Lillard, *The Great Forest*, p. 335; Horn, *This Fascinating Lumber Business*, p. 51.

Logging crew at Doucette, with woods boss Peter Doucette. *Courtesy of Stephen F. Austin State University, Forest History Collections.*

sort of open range on which their stock could graze at will.[32]

Actually, such seasonal woods burning did little damage to the lordly stands of virgin longleaf pine and, by eliminating the accumulated debris, often prevented much more destructive fires during the late summer, when the forests were tinder-dry. In promoting spring grass and a clean forest floor, however, the fires also destroyed the longleaf seedlings and young pines. With the destruction of the virgin forests the second-growth forests presented an entirely different problem, and fires unsupervised and uncontrolled could cause them great damage.[33]

[32] Stewart H. Holbrook, *Burning an Empire*, pp. 170–71; Wahlenberg, *Longleaf Pine*, pp. 142–43.
[33] Wahlenberg, *Longleaf Pine*, pp. 144–54; A. A. Brown and A. D. Folweiler, *Fire in the Forests of the United States*, pp. 38–41. U.S. Department of Agriculture, *Trees: The Yearbook of Agriculture*, 1949, pp. 517–27.

Much damage was also done by malicious burning. A later survey disclosed that 80 percent of the nation's forest fires occurred in the South, and it can be safely assumed that Louisiana and Texas had their share of such fires. It has been variously estimated that 35 to 50 percent of the fires were set by arsonists and "fire bugs" suffering from pyromania. Some fire bugs simply wanted to spite the "big man," to "whittle him down to size." Others set fires because they liked to smell the pine smoke; still others, merely because they enjoyed the excitement. Natives felt wronged when a lumber company, especially a large absentee owner, tried to fence their domain or ban hunting. Some sought retaliation when progressive foresters introduced the practice of killing scrub oaks and other low-grade hardwood trees to allow the young pines to gain dominance and achieve a more rapid growth. An incidental effect of "deadening," or killing, the hard-woods was to drive off the game, especially squirrels, by destroying the source of the acorns and nuts that were their food supply. This was certain to arouse the ire of a native hunter. In revenge, he would set fires in the pine forests that might consume thousands of acres of valuable timber. A ditty that became current before the Great Depression expressed the challenge of the nesters, hunters, and local farmers:

You've got the money,
We've got the time,
You deaden the hardwoods
and we'll burn the pine.[34]

Rivaling the damage caused by forest fires were the losses caused by rooting razorback hogs. The East Texas or western Louisiana "piney woods rooter" is a lean, long-legged, re-sourceful forager, and local farmers had long felt that their livestock, including the razor-backs, had an inalienable right to graze on the "public domain," including the forest pasture-lands of the great lumber companies. The hogs were especially damaging to young pine seed-lings, seemingly finding the cortex of the main taproot particularly luscious. They also ate the foliage, the bark, and the plant itself, as well as trampling and rooting out the general area. Again, as the lumber owners and timber mana-gers discovered, any attempt to fence the area or protect their property with no-trespass signs was regarded as virtually a declaration of war by the local population. As a result, in the Texas-Louisiana region cattle and hogs contin-ued to graze, and natives continued to fish and hunt in the great forests of the Piney Woods despite the advent of the great lumber com-panies and the logging bonanza that followed.[35]

Most members of the woods crew insisted to strangers that they were loggers rather than sawmill workers. Despite their dependence on the mill for their employment and livelihood, most loggers exhibited hostility to the mill. They often declared that they would not work in the mill; it was too confining, too noisy, too dirty. They seemingly rejoiced in the open air, the varieties of nature, and the constant chal-lenge of the great forest. To most logging work-ers the many and varied tasks of the woods crew were much to be preferred to the same-ness and monotony of a job at the mill.[36]

Loggers preferred life in the woods despite the fact that logging was one of the most haz-ardous occupations in the whole field of man-ufacturing. According to the United States Department of Labor, logging was more hazardous than sawmilling and, indeed, seven times as dangerous as the average of all manu-facturing industries.[37] A compilation of log-ging accidents at one large lumbering com-pany included seventeen serious, lost-time accidents in the woods in one year. The list in-cluded such injuries as being cut by an ax, being struck on the foot or leg by a falling limb or tree, being struck by a rehaul (skidder) line, and being struck by tongs. The injuries pro-duced an array of cuts, fractures, mashed limbs, sprains, and bruises and an average time lost per injury of twenty-two days.[38]

In a more comprehensive survey of logging accidents the *Gulf Coast Lumberman* ana-lyzed 1,044 accidents, including 11 fatalities, and charged that most were preventable and had cost the industry more than $250,000, not counting loss of earning power to the injured and his family. Injuries were caused by falling

[34] Conversation with Clyde Woodward, Sr., 1958; conversa-tion with Needham B. Weatherford, 1959. See also Ed Kerr, "Southerners Who Set the Woods on Fire," *Harper's* 217 (July, 1958):28–33.

[35] Conversation with N. B. Weatherford, 1959; Wahlenberg, *Longleaf Pine*, pp. 178–79.

[36] Conversations with George Morrison, Angelina County Lumber Company, August 6, 1958; Edward Harper, Kirby Lumber Company, August 5, 1958; Charles Welch, W. T. Carter and Brother Lumber Company, August 5, 1958.

[37] Allen, *East Texas Lumber Workers*, p. 107, quoted from *Occupational Hazards for Young Workers*, U.S. Department of Labor, Children's Bureau Bulletin no. 276, report no. 5, Log-ging and Sawmilling Industries Series (1942), p. 21.

[38] Angelina County Lumber Company, Workman's Compen-sation Reports for 1925, ACLCo. Papers.

objects, cables, and falls; accidents in which workers were caught between logs or in machinery; overtaxation (sprained backs, hernias, and so on); and injuries from axes, saws, grabhooks, and other tools.[39]

Some of the men logging in the pine forests of the Southwest were independent farmers who grew crops in the spring and summer and then took jobs, perhaps with their mule teams, in the woods for the fall and winter. Many in this group were contract loggers who cut and sold logs, sometimes from their own lands. But the great majority of logging workers were regular employees of one of the large companies, living in the company town, either at the mill or at the front, and traveled to and from the mill on the company railroad. To many, perhaps most, of them it was a career and a way of life rather than merely a temporary job.[40]

In the lumbering operations of Texas and Louisiana both blacks and whites worked in the woods. In the industry as a whole the proportion was roughly two-thirds white and one-third black, and the ratio was approximately the same in logging, as distinguished from the sawmill. In most companies members of the two races rode to the woods together, worked together, and returned home at night together but did not eat together or live together. In mill towns and logging camps the two races occupied separate sections. The great majority of blacks were laborers, and in most companies whites held all the skilled and supervisory jobs. Nearly all workers of both races were natives of the Piney Woods, and a majority had lived most of their lives in the county in which they were employed. A few immigrants were engaged in logging operations; some, like Peter Doucette, from Canada, in a supervisory capacity. Many companies along the Gulf Coast employed some Mexican laborers, and one East Texas firm, W. T. Carter and Brother, regularly employed a few Indians from the Alabama and Coushatta Reservation in its woods crew. By and large, however, logging in East Texas and the Gulf Coast was carried on by native-born Americans, white and black, who were born and bred in the Piney Woods, were at home there, and, according to their own testimony, would not live anywhere else.

Piney Woods loggers worked hard. The hours were long; the work was exhausting and dangerous; the pay, by any contemporary standard, was low; and the fringe benefits were minimal. On the job they were cheerful, competent, wasteful, and at times careless, but they were also purposeful. Their goal was to supply logs to the "big mill"—if possible, to swamp it in logs. With this goal in mind the loggers cut down the great pine forest with no thought of the future and little of themselves. If they could exceed their quotas, they were content.

[39] "Mayhem in the Woods," *Gulf Coast Lumberman*, June 9, 1949, p. 36.
[40] Conversation with H. Z. Collier, Houston County, July 29, 1964; conversation with Needham B. Weatherford, October 13, 1959.

Sawmill, millpond, and log train framed against a   *State University, Forest History Collections.*
cloudy Texas sky. *Courtesy of Stephen F. Austin*

CHAPTER 5

At the Mill

Good timber, sawn slightly full both ways
. . .—inspection report, Texas sawmill,
1908

To the visitor the sawmill was a strange, interesting complex. To the worker the mill was a monster—a great, glowering, omnivorous, insatiable, omnipotent monster. It devoured the men, father and son; it ate up the forests; it transformed the countryside into a desert of sawdust dunes; it destroyed the tranquillity of rural life; and finally, more often than not, it destroyed itself—by fire. Sawmill work offered long hours, low pay, little chance of advancement, an uncertain future, and, by the law of averages, a good chance of at least one serious injury.

In most sawmill towns the mill dominated not only the landscape but also the economy. It was often the only place of employment. In a region of little hard money mill wages were often the only source of cash. In most mill towns the sawmill represented a way of life, and frequently two or even three generations worked in turn in the mill.

The standard equipment of the big mills appears to have been the double-band mill—two band saws on separate log carriage lines with or without a gang saw. This mill, firmly set on solid foundations and in a well-constructed building, came equipped with edgers, trimmers, dry kilns, planers, and yard sheds. There were variations on this equipment. Here and there a mill operated with a pair of giant circular saws. The Central Coal and Coke Company mill in Houston County, Texas, boasted of a triple-band rig, three log carriages, and a gang saw. The Great Southern Lumber Company, at Bogalusa, Louisiana, had four eight-foot band saws that ran night and day.[1]

The big mills, however, represented only a small minority of the sawmills in operation in the area at any given time. The *Texas Almanac, 1910*, estimated that there were "about 625 lumber mills in the state and of these about 375 to 400 are of a size to be of importance. Of this latter class the big mills made up but a small percentage."[2] There were probably never as many as 100 big mills in Texas, as opposed to 200 or 300 permanent to semipermanent small-to-medium-size mills and as many more small portable sawmills. The last were crude, simple, often powered by a tractor or a model T Ford engine, and wasteful, owing to their wide kerf. Larger manufacturers often referred to them as "peckerwood" mills. They usually cut only low-quality green lumber that their owners sold locally or to a larger mill for

[1] Conversation with H. Z. Collier, Ratcliff, July 31, 1964; conversations with Clyde J. Woodward, Sr., 1958–1963, Oral History Collections, Forest History Collections, Stephen F. Austin State University Library, Nacogdoches. Unless otherwise noted, interviews and conversations cited are in these collections, and all interviews were conducted by principal author Maxwell. See also Ed Kerr, *History of Forestry in Louisiana*, p. 3; James Boyd, "Fifty Years in the Southern Pine Industry," *Southern Lumberman* 144 (December 15, 1931): 59–67.

[2] *Texas Almanac, 1910*, p. 112. Big mills were those with a capacity of 80,000 to 100,000 board feet and more a day. Small mills were those with a capacity of less than 30,000 board feet a day. Middle-sized mills were those with a capacity of 30,000 to 80,000 board feet a day.

Large circular saw. *Courtesy of Forest History Society.*

remanufacture. Farmers, ranchers, and others operated such portable mills part-time, setting them up when there was a demand for lumber or when there was a slack time in their regular jobs. They were often literally here today and gone tomorrow.[3]

Most of the lumber was cut by the big, permanent mills. They employed the vast majority of the workers, built the company towns, formed the trade organizations, engaged in interstate and overseas traffic, and spoke for the industry. As had been indicated, the big mills and their owners were the heart of the lumbering saga in the Gulf South.

The progress of a sawmill from small-mill to big-mill category was illustrated by the story of the Thompson family. In 1852, John Martin Thompson, with his brothers and later his sons, built a small sash-type sawmill south of Kilgore, in northern Rusk County. It had a capacity of about 2,000 board feet a day. It burned in 1853, and Thompson built a larger mill with a circular saw that increased daily production to about 5,000 board feet. During the next twenty-five years he built a series of mills in the same general location, each one

Large circular saw showing removable teeth. *Courtesy of Forest History Society.*

somewhat larger and more efficient than the last. Thompson and his brothers supplied the local community with lumber and even hauled finished lumber as far as Dallas and Fort Worth by wagon and mule team.[4]

In 1881, the Thompsons decided to move their operations to Trinity County, where they could exploit the great stands of longleaf pine that up to that time had remained largely untouched. They located on the Trinity and Sabine Railroad, which was still under construction, and rebuilt their mill at a settlement called Willard. In 1888 they built a new mill, featuring a large circular saw, a top saw, an edger, trimmers, and a gang saw, which had a capacity of about 80,000 board feet a day and moved the Thompsons into the group of big-mill operators.

[3] Conversation with Needham B. Weatherford, W. T. Carter and Brother Lumber Company, 1958; *Texas Almanac, 1926.* Small portables were also called "ground rattlers" or "coffeepots." Stanley F. Horn, *This Fascinating Lumber Business,* pp. 141–43.

[4] *American Lumberman,* July 17, 1902; September 26, 1908; conversations with Hoxie H. Thompson, 1958, 1959; conversations with Mrs. Hoxie H. Thompson, 1959, 1962, 1963.

A millpond. *Courtesy of Stephen F. Austin State University, Forest History Collections.*

From there the Thompsons expanded their operations to other East Texas points. John Martin Thompson increasingly turned over the daily operations of the family business to his several sons, who established new plants at Doucette, Grayburg, and Trinity under various names. When the timber at Willard was exhausted, Thompson moved to a new location on the Houston, East and West Texas Railway, which he called New Willard. To coordinate these enterprises, J. Lewis Thompson, an older son, established executive offices in Houston so that he could oversee operations and supervise purchases from a central base. Any one of the several mills that the Thompson family built after 1900 could be considered typical of the big mills operating in Texas before World War I.

Any description of the operation of a major sawmill must begin at the millpond. There the loggers dumped their products after the long haul from the woods by railroad, wagon, and later, truck. Most companies of the bonanza period considered the log pond an essential part of their operations, and if none already existed, they built one. A typical log pond covered several acres and provided storage for one to four million board feet of logs. Log-pond storage had a number of advantages. It was easy to dump the logs into the pond from the railroad log cars with a minimum of handling. Since most logs were skidded by either mules or power skidders, they were usually dirty and grit-filled when they arrived. The pond washed them off, thus reducing damage to or dulling of the saws. Pine logs float, but with most of the surface underwater, where they are less exposed to various bark beetles and other insects. From the pond workmen could easily steer the logs onto the jack ladder (an endless

A load of longleaf pine ready for the mill. *Courtesy of Stephen F. Austin State University, Forest History Collections.*

conveyor chain equipped with lugs), on which they were hauled up to the log deck.[5] At a mill that had no log pond, the railroad-dock workers simply piled the logs on a siding or log dock, to be transferred to the mill as needed.

Most large sawmills were two- or three-story buildings with the main saw machinery on the second floor. The jack ladder raised the logs from the pond to the log deck on the main floor. There activity began in earnest. The scaler measured and recorded the estimated size of the log in board feet. He also operated the cutoff saw, sawing long timber into efficient cutting lengths. Then he operated the log kicker, which forced the log off the trough and onto the sloping deck. In the typical mill there was a double log kicker, and the scaler sent the logs alternately right and left to the band saws at either side of the mill (there were often a long side and a short side—frequently one log carriage was specially built to handle extra-long logs; the other was called the short side). The scaler was responsible for keeping an adequate supply of logs on the log deck. If production was interrupted for any cause, he stopped the jack ladder so that logs would not pile up.[6]

From here the sawyer took charge. He controlled the log loader and deck stops that loaded one log onto the carriage, stopped and held the next log in reserve at the edge of the deck, and then in turn loaded it. Once the log

[5] An additional advantage of having a log pond at the sawmill site was the availability of a water supply in case of an emergency, such as a mill fire. Conversation with Needham B. Weatherford, July 23–25, 1958.

[6] Nelson Courtlandt Brown, *Lumber: Manufacture, Conditioning, Grading, Distributing, and Use,* pp. 40–49.

was on the carriage, the sawyer adjusted, turned, and blocked it in preparation for its first pass through the saw.[7]

The carriage was essentially a small railroad car powered by steam, by either cable or piston action, set on tracks, and fitted out with the necessary setworks, blocks, dogs, guides and buffers. The carriage must be properly adjusted and aligned and must function properly and accurately. In the early days four men rode the carriage and set the blocks by hand. Two other workers with cant hooks turned and shifted the log on command at the end of each run. A series of inventions between 1890 and 1915 reduced the block setters to two and then to one, who operated a number of controls powered by compressed air to regulate the blocks, dogs, and knees as directed. Likewise the log turners on the floor were replaced in time by a powerful steam-driven inverted gothic T bar with teeth that could turn or flip the largest log as easily as a boy could toss a baseball bat. Workers called it a "steam nigger."[8]

Many men took part in the inventions that revolutionized the log-handling machinery at the saw mill. In Lufkin, Texas, W. C. Trout, a practical mechanic, invented and patented setworks that bear his name and were widely used in the industry, especially in Texas and other parts of the Southwest. In Clinton, Iowa, Chancy Lamb, an inventor and mechanical wizard, devised many improvements, including a log trimmer, a steam-powered log turner, and an edger with movable blades. The large manufacturers of sawmill equipment, such as Filer and Stowell of Milwaukee, constantly improved their products by acquiring patent rights to new inventions. Visitors at the world's fairs in Chicago in 1893 and Saint Louis in 1904 were amazed and fascinated by exhibits of advanced automatic sawmill machinery.[9]

High-speed band saw showing setworks. *Courtesy of Forest History Society.*

Not all Texas lumbermen were eager to install automatic machinery. The story was long current in East Texas that Aubrey Carter, the son of the founder of W. T. Carter and Brother Lumber Company and for many years its president, refused for a time to purchase one of the new oscillating steam log turners which would replace several workers. Pressed to make the change by one of his fellow lumbermen, Carter is reported to have replied, "Hell, no! It wouldn't trade at the commissary."[10]

The most noteworthy feature of the big double-band mill was the bandsaw itself. This endless ribbon of highly tempered steel, usually twelve to eighteen inches wide and thirty to fifty feet long, operated on two huge wheels placed in line above and below the cutting area. Because of its narrow kerf, often little thicker than a knife blade, and its great accuracy in cutting, the bandsaw gradually replaced the circular saw at the larger mills. Credit for the development of the band saw in the United States is usually given to J. R. Hoffman, an Indiana lumberman, who began experimenting with the idea as early as 1860.

[7] Ibid., pp. 30–51.

[8] Ibid., pp. 55, 59–64; conversation with Ed Bird, W. T. Carter and Brother Lumber Company, July 25, 1958.

[9] Conversation with A. E. Cudlipp, Lufkin Foundry Company, March 8, 1960; conversation with Simon W. Henderson,

Jr., Angelina County Lumber Company, August 10, 1964; H. W. Whited, Frost-Johnson Lumber Company to Filer and Stowell Company, May 17, 1910, in *American Lumberman*, January 16, 1915, p. 93; Richard G. Lillard, *The Great Forest*, p. 154.

[10] Conversation with F. I. Tucker, Nacogdoches, 1960.

Band saw in operation. *Courtesy of Forest History Society.*

Logs on the deck, ready for the sawyer's signal. *Courtesy of Stephen F. Austin State University, Forest History Collections, Thompson photo albums.*

After many failures he obtained suitable steel from abroad and operated a successful band mill in Fort Wayne for many years. Band saws came into general use in the upper Mississippi Valley sometime in the 1880s, and evidence exists of their use in the Great Lakes region before 1890. Just when the band saw was introduced into the Southwest is uncertain, but it was probably not much earlier than 1891. By that time Lutcher and Moore, and perhaps other operators, had installed band saws.[11]

The key man in the entire sawmill operation was the sawyer. He had a good knowledge of lumber and the essentials of grading and was able to size up a log in a few seconds to determine how best to saw it to produce the most and the highest-quality lumber. He was one of the most experienced men in the mill and one of the best paid. The sawyer and his block set-

ter operated as a team. Because of the noise in the mill the sawyer gave his directions by means of hand signals, and the setter, who alertly kept his eyes on the sawyer, executed the orders with smoothness and dispatch. The sawyer controlled not only the release of the log to the carriage and its adjustment but also the shuttling back and forth of the log against the headsaw. As he cut the log into boards, he usually had to rotate the log frequently to obtain the best grain and the highest grade. In mills equipped with gang saws, the sawyer also must decide whether to complete the sawing at the headsaw or to pass the "cant" (a log from which one or more slabs have been removed) onto the gang saw. All double-band mills had at least three experienced sawyers; most had four or five. In many mills the manager himself had had sawyer experience.[12]

As the boards fell free of the log and carriage after passing the headsaw, the worker called the "tail sawyer" (or "off-bearer") guided them onto a conveyor system. In the earlier mills this was simply a series of rollers over which workers pulled or pushed the boards with hooks. Modern mills have systems of "live" (powered) rollers that carry the board to its next station. The conveyor detoured the

[11] A band saw was invented in England in 1808 by William Newberry but seems not to have been put to general use. Apparently Hoffman developed his idea independently. Brown, *Lumber*, pp. 72–75; Horn, *This Fascinating Lumber Business*, p. 137; Robert F. Fries, *Empire in Pine: The Story of Lumbering in Wisconsin*, pp. 62–63; Ralph W. Hidy, Frank E. Hill, and Allan Nevins, *Timber and Men: The Weyerhaeuser Story*, pp. 165–68. Most band saws used in Texas and Louisiana were single cutting bands, that is, with cutting teeth on only one side. Some mills had double cutting saws, with teeth on both sides, but such a saw necessitated a more complex arrangement to take care of the lumber cut on the return trip of the carriage. Brown, *Lumber*, p. 78; F. H. Gilman, "History of the Development of Sawmill and Woodworking Machinery," *Mississippi Valley Lumberman* 36 (February 1, 1895): 63.

[12] Conversation with Ed Bird, July 25, 1958; Horn, *This Fascinating Lumber Business*, pp. 144–46.

Gang saw at work. *Courtesy of Forest History Society.*

Trimming operation at the mill. *Courtesy of Stephen F. Austin State University, Forest History Collections, Thompson photo albums.*

large cants to the gang saw, where they were sawed into identical boards, generally 2-by-6 or 2-by-8 boards, at one time. The advantage of the gang saw was that it could produce an enormous quantity of board feet of lumber an hour. Its disadvantage was its lack of flexibility.

Single boards went to the edger. The edger-man also must be experienced in judging lumber, for it was his job to remove the "wane" (rounded edges) from the boards, divide them into boards of standard width, and separate the clear boards from the knotty, ragged, or otherwise lower-grade boards.[13]

Sawmillers also on occasion installed "resaws," either horizontal or vertical band saws, to increase production. Resaw bands were generally smaller than headsaws, and the steel was of a thinner gauge. By this method the sawyer at the main band saw reduced the log only to heavy cants and passed them onto the resaw. It was estimated that a mill could increase production as much as 30 percent by use of a resaw. Mills equipped with resaws did not usually have gang saws. The resaw was positioned behind the headsaw and ahead of the edger.[14]

After the lumber had passed the edger, the conveyor belt carried boards, slabs, odd

lengths, and trash alike toward the trimmer and stackers. Along the way it passed the man in the "bear pit," who separated merchantable boards and pieces from the slabs and waste lumber and sent each in its proper direction: the good lumber to the end of the mill and the rest to the "hog and slasher" (a power-grinding machine), to be ground up into fuel or waste. The man in the bear pit was one of the unsung heroes of the sawmill. There he stood, waist deep in the pit, with the lumber from the edger coming at him in torrents. He must either clear it or be covered up. Paid only as a common laborer—and in Texas and Louisiana usually a black man—he was not rated one of the key workers in the sawmill. Yet his task required physical strength, quick eyes and hands, and at least an elementary knowledge of what was merchantable and reclaimable lumber. The man in the bear pit had little time for musing, for he worked in silent desperation. A daydreamer would stop the mill.[15]

From the bear pit the lumber passed to the end of the mill, where it went through the trimmer before leaving the mill. The trimmer consisted of a series of small circular saws bal-

[13] Brown, *Lumber*, pp. 93–100; Horn, *This Fascinating Lumber Business*, p. 147.
[14] Brown, *Lumber*, pp. 104–107.

[15] Conversation with Ed Bird, July 23, 1958. In the early-day sawmill the conveyors were arranged so that the output of both saws passed over a single bear-pit position. In the larger mill the bear-pit worker cleared for only one saw and edger. Even so he was truly "fighting the bear."

Saw filer's room. The task of keeping the band saws sharp was a key job. *Courtesy of Stephen F. Austin State University, Forest History Collections, Thompson photo albums.*

anced on a set of overhead weights above an endless steel belt along which the boards moved at the right angles to their previous path. Above them the operator of the trimmer—also called the trimmer—sat or stood in a cage, operating keys or levers that raised and lowered individual saws that cut the boards into the desired lengths. The trimmer was one of the key men in the mill, for he must be an expert lumber grader who could quickly decide whether a given board should be cut shorter to eliminate a defect and raise the grade or simply squared off at the ends and sent through as a standard length. In the more sophisticated modern mills the trimmer sits in his chair in the operating cage controlling the saws with a keyboard much like that of an organ.[16]

High above the headsaws on the third floor of the mill was the saw filer's room. To many owners the saw filer, not the sawyer, was the most important single person in the mill. He often was the highest-paid worker and was perhaps the most difficult to replace. It was not unusual for a saw filer to spend twenty or

thirty years on one job, during which time he might train two or three apprentices. The filer who could keep the many saws razor-sharp and clean; make certain that their mountings were strong, solid, and plumb; and keep ahead of the constant demands for replacements did much to ensure that the mill ran efficiently and profitably.

On an average the mill manager replaced the headsaws twice a day, the saws in the gang saws once, and the edger saw blades once. The others, including the trimmer, were replaced less often. If the sawyer encountered logs with foreign substances, such as wire or spikes, embedded in them, he might find himself spending most of the day changing saws. The manager preferred to change headsaws at the noon break and after quitting time at night. Since saws were among the most expensive items in the sawmill, a skilled and contented filer was one of the best investments that a millowner could make to ensure that shutdown time was kept to a minimum.[17]

The saws' footings and other machinery of the mill were beneath the main floor of the sawmill proper. The power plant—boilers, engines, and giant flywheels—was housed in a separate building nearby. The power requirements for a large double-band sawmill with several power features and auxiliary saws were high, often 1,000 horsepower or more. Fuel for the boilers was usually provided from the waste from the sawmill—slabs, chips, and sawdust that were fed into the firebox by blowers or chain conveyors or, in older mills, by hand.[18]

In the production of lumber there was enormous wastage, estimated as high as 40 percent of the round log. The use of marginal materials and the disposal of the rest provided continu-

---

[16] Horn, *This Fascinating Lumber Business*, pp. 147–48; Brown, *Lumber*, pp. 100–104; conversation with Ed Bird, July 23, 1958.

[17] Conversation with Simon W. Henderson, Jr.; conversation with T. L. Wilson, Angelina County Lumber Company, August 10, 1954; Brown, *Lumber*, pp. 82–85. See also Ralph W. Andrews, *This Was Sawmilling*, pp. 45–51.

[18] *American Lumberman*, September 26, 1908, pp. 109–10; Brown, *Lumber*, p. 113; Horn, *This Fascinating Lumber Business*, pp. 142–43.

Scene at mill yard showing stacked lumber and waste conveyor. *Courtesy of Stephen F. Austin* *State University, Forest History Collections.*

ing problems for millowners and engineers. Conveyors carried the rejected pieces from the bear pit and the trimmer to the lower floor, where short lengths and strips were salvaged for laths, moldings, crates, boxes, and similar products. The rest was fed into the hog, a machine that reduced waste into small chips to be used for fuel. The blower system carried the chipped fuel to the boilers, where it burned much more uniformly and more efficiently than "unhogged" waste. Most of the larger mills installed hogs in the decade after World War I. Even earlier millowners had become concerned with the use of marginal pieces that had previously been discarded as waste.[19]

In all large sawmills far more sawdust was produced than the mill could use as fuel. Efficient and safe disposal of sawdust was a continuing problem for millowners. The usual method was to use a conveyor belt or blower system to carry it well away from the sawmill

[19]Brown, *Lumber*, pp. 110–12; *American Lumberman*, September 26, 1908; conversation with Ed Bird, July 23, 1958.

General view of Trinity County Lumber Company, Groveton, ca. 1916, with stacked lumber in fore-ground. *Courtesy of Stephen F. Austin State University, Forest History Collections.*

operation and pile it up. Local residents who wanted sawdust for fuel or other purposes were welcome to it, and periodically it burned, either accidentally or by design. A burning or smoldering sawdust pile was a characteristic sight in Texas sawmill towns during the bonanza period.[20]

From the trimmer the lumber emerged into the rough-lumber yard, and there workers sorted it according to widths and lengths. In

the early days of the big mill the task was laborious and time-consuming and required a large number of workers. After sorting the "green" lumber into proper piles, the workers loaded it into wagons or carts and moved it to the drying area, where they stacked it to allow it to dry. It was a slow process, often requiring weeks and frequently resulting in sap-stained lumber. Various experiments were undertaken to try to hasten the drying process, and several kilns were devised. The first kiln introduced into Texas was probably the so-called Arkansas smoke kiln. In this crude, boxlike structure the lumber was piled on a platform, and a fire was built underneath. The heat and smoke

[20] Conversation with Ed Bird, July 23, 1958; conversation with Clyde J. Woodward, Sr., 1962. This description applied to most big mills from the turn of the century to the 1950s. The introduction of the log debarker made the processing of chips for sale to pulp and paper mills an important additional source of income.

Planing mill showing machinery for dressing lumber. *Courtesy of Stephen F. Austin State University,* *Forest History Collections.*

were drawn upward through a hole in the roof. By this means some of the moisture was removed from the lumber. As one writer commented, the smoke kiln was very unreliable: "About the only thing . . . which could be depended upon was the assurance that, sooner or later, it would set fire to the lumber and probably burn down the entire operation." [21]

Gradually new developments improved the

smoke kiln, and before the turn of the century a number of firms were offering effective brick and steel dry kilns. [22] Most of these kilns employed the same general process to remove moisture, using steam coils and fans or blowers inside the kilns. By carefully regulating the moisture and heat in such a kiln the operators

[21] Horn, *This Fascinating Lumber Business,* p. 150.

[22] F. H. Gilman, "History of the Development of Sawmill and Woodworking Machinery," *Mississippi Valley Lumberman,* February 1, 1895, p. 67.

Planing mill showing planer and matcher machines. *Courtesy of Stephen F. Austin State University, Forest History Collections, Thompson photo albums.*

were able to dry and season lumber in four to six days, whereas it might take as much as a month by the air-drying method. If the lumber was properly stacked on a dolly before it entered the kiln and the dollyway tracks extended through the kiln, the lumber could be processed from the "green" lumber-sorting tables through the kiln to the rough-seasoned lumber shed without rehandling. By 1900 most of the big mills in Louisiana and Texas followed this or a similar procedure in their stacking and drying yards.[23]

From the yards and the dry kiln the lumber went to the planer for remanufacture. In the early days of lumbering, both in Texas and elsewhere, the operators sold their products in the rough. The purchasers dressed and finished the lumber to suit their purpose. Beginning about 1880, however (about the time Lutcher and Moore came to Texas), lumber manufacturers dressed only boards, and those on only two sides. After the turn of the century it became customary to dress both boards and larger dimension lumber on all four sides. Larger mills also equipped themselves to provide various finished products, such as flooring, ceiling, siding, partition, and molding.[24]

The large millowner found a number of advantages in dressing his own lumber and remanufacturing it to meet customer demand. An important advantage was the reduction of shipping weight. Freight rates were based on the weight per 1,000 board feet. By finishing the lumber, the manufacturer could reduce the weight by about 20 percent while increasing its value. Remanufacture also enabled the lumberman to achieve a closer use of rough material at the plant by meeting the demands for certain finished products. Moreover, the manufacturer of finished products could sell directly to country retail yards, where there were both less competition and a market for a wider variety of products than would be called for in large central markets. These advantages accrued only to the larger manufacturers, because only the big mills had sufficient volume to justify the purchase of planing machinery.[25]

The essential item of equipment in the planing mill was the planer and matcher. This machine was equipped with both round cutting cylinders and sideheads, was mounted on a sturdy steel floor or bed, and was designed to operate at high speeds. Additional machines were added as the volume of production increased, and it was not unusual to find new planers and matchers operating side by side with machines that were twenty or thirty years old, each doing the same job with relatively equal efficiency. By means of a profile attachment the planer and matcher could produce tongue-and-groove flooring, ceiling, and other special items. The mill often had a molder, which was essentially a smaller, lighter model of the planer. There were often also vertical or horizontal resaws, ripsaws, and cutoff saws.[26]

---

[23] Brown, *Lumber*, pp. 146–52; conversation with Ed Bird, July 23, 1958.

[24] *American Lumberman*, January 18, 1908; J. Richards, "A

Treatise on the Construction and Operation of Woodworking Machines," *Journal of Forest History* 9, no. 4 (January, 1966): 16–23.

[25] The Southern Pine Association estimated in 1947 that only about 10 percent of some fifteen thousand sawmills in the entire South were equipped with proper planing and grading facilities. Brown, *Lumber*, pp. 157–59; Ralph C. Bryant, *Lumber: Its Manufacturing and Distribution*, pp. 240–49; Horn, *This Fascinating Lumber Business*, p. 152.

[26] Conversation with Simon W. Henderson, Keltys, July 9,

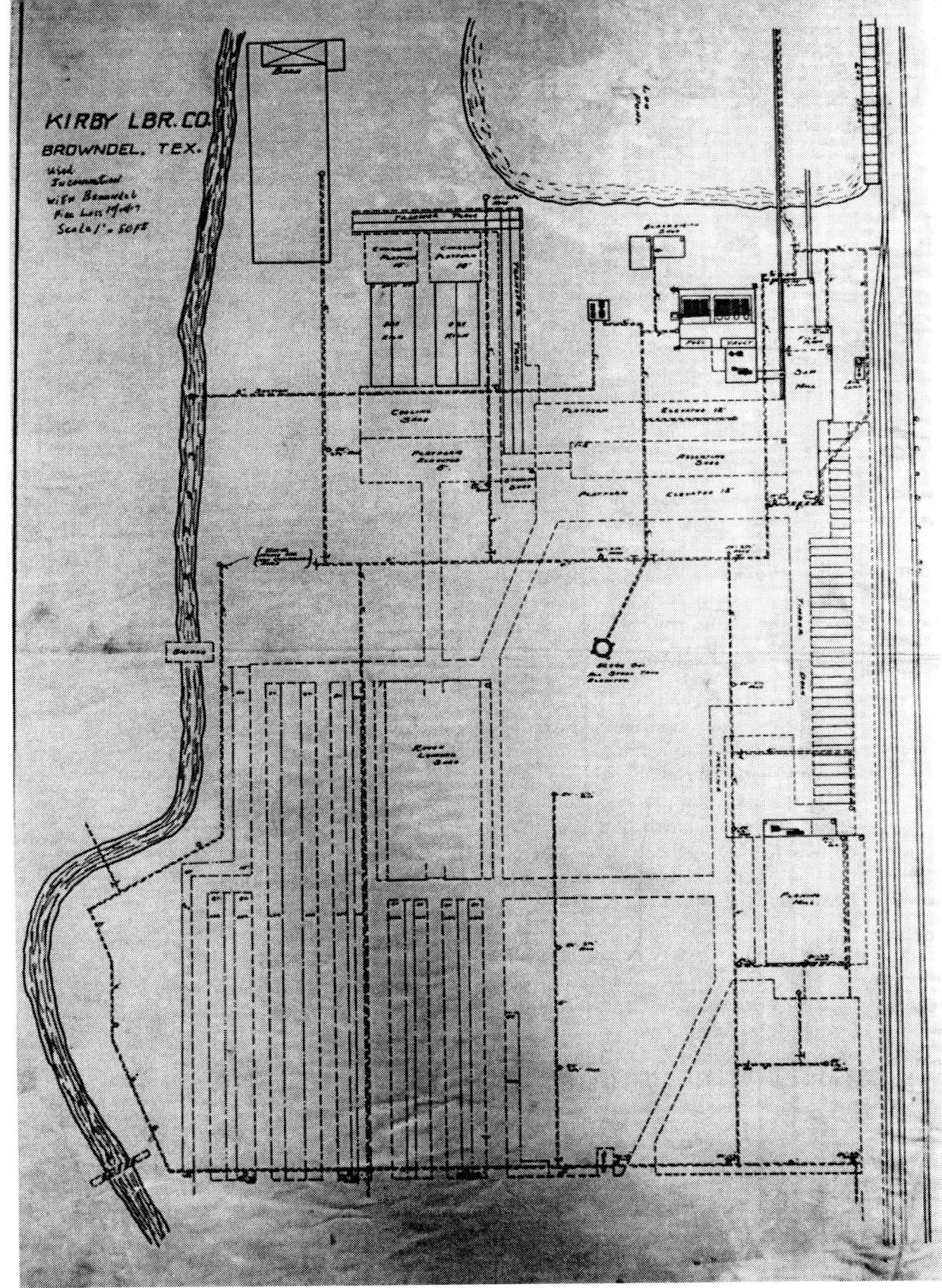

Ground plan of Kirby Lumber Company sawmill at
Browndell, 1917. *Authors' collections.*

The development of planing machinery has had an interesting history. From time immemorial carpenters had smoothed rough boards with a hand plane, which is hard, slow, backbreaking work. William Woodworth, a carpenter in upstate New York, invented the first crude machine planer in 1828. He did little to develop his invention and later sold it to a group of promoters, who proceeded to lease the machine at a royalty based on the amount of lumber processed and thus made the planer very profitable. The United States Congress extended the patent monopoly on the planer, and it was not until 1856 that the patent was at last terminated and the planer became available for competitive manufacture.[27] In the following years the planer was constantly improved until by the turn of the century it had been developed into the efficient machine that is still in general use today.

The planing mill traditionally also has manufactured another, less attractive, product. Every planing mill had its quota of "three-fingered Willies" who had suffered hand injuries while working at the planing machines. Between 1922 and 1926, according to the National Council on Compensation Insurance, planing-mill accidents accounted for 10 percent of all the accidents in the entire logging and lumbering industry. The lumber manufacturers were reluctant to install guards on their woodworking machinery, and the states (particularly Texas) were slow to require them to do so. Until after the passage of the Texas Workmen's Compensation Act of 1913, there appeared to be little interest in the problem;

the danger was assumed to be one of the hazards that went with the job and one for which the employees voluntarily assumed the risk as a condition of employment.[28]

Because of the fire danger posed by accumulated shavings and refuse, most mill operations had separate power plants for the planing mill. Most of the older plants used steam-power plants, similar to those employed in the main sawmills, only smaller. If a fire or an accident occurred in the planing mill, it would not force the entire mill to shut down. Increasingly in the 1920s and later, millowners converted their power plants to electrically driven motors, which proved to be both more efficient and more economical.[29]

The heart of every lumbering operation was the sawmill. The lowering smoke from the tall stacks was the dominant feature on the skyline of a hundred mill towns. When the mill rested, the people slept. When its whistle blew, the workers sprang to life. When the mill ran, all was well: the men worked, the owners made a profit, and the mill produced finished lumber, which was the goal of the entire operation.

During the heyday of Texas lumbering the typical permanent mill manufactured 20 to 75 million board feet of lumber a year. The finished product, collectively a significant fraction of the lumber output of the United States, was ready to market locally, nationally, or abroad. From that point the success of the owner-manager in marketing his product determined the success of his venture.

1962; Horn, *This Fascinating Lumber Business*, pp. 152–53; Brown, *Lumber*, pp. 167–72; Bryant, *Lumber*, pp. 251–57.

[27] At the expiration of Woodworth's patent Joseph P. Woodbury obtained a patent on a improved planer and organized a company to perpetuate monopoly over this type of machine. The patent, however, was nullified in a spectacular case (*Woodbury Patent Planing Machines Company* v. *Allen W. Keith*) in which defending manufacturers banded together to prove that the main features of Woodbury's machine antedated his patent application. F. H. Gilman, "History of the Development of Sawmill and Woodworking Machinery," *Mississippi Valley Lumberman* 26 (February 1, 1895): 59–69. See also Richards, "A Treatise on the Construction and Operation of Woodworking Machines," pp. 17–23.

[28] D. D. Smith, secretary, National Council on Compensation Insurance, to author Maxwell, October 15, 1959.

[29] Conversation with Simon W. Henderson, July 19, 1962.

CHAPTER 6

Marketing

If you ever get pine splinters in your hands,
you never get them out.—Claude W. Bryant

IN the earliest days the sawmill operator usually sold his products at the mill, often operating intermittently, sawing a run of lumber only to fill a specific order. Before the Texas Revolution, William P. and Robert Harris operated a lumberyard in connection with their sawmill on Buffalo Bayou and shipped lumber to Galveston for sale. As early as 1829, Peter Ellis Bean maintained a lumberyard in Nacogdoches that was at least a mile or two from his mill on Carrizo Creek. Robert Jackson had most of his stock swept away in a spring flood on the Sabine, where he had his mill and yard in 1847.[1] In this early period lumber was scarce, and its manufacture was a slow, laborious task with a water-powered sash-saw mill producing only one to two thousand board feet a day. Prices were high, even for green, undressed boards, and transportation costs frequently doubled the price of purchased lumber. Under these conditions separate marketing organizations were unnecessary, and sales techniques were primitive. The millowner could market all the lumber that he could produce, and, according to early travelers, there was a constant shortage of lumber in Texas despite the abundant pines.[2]

The development of large mills in the two decades after 1880, particularly the introduction of the band saw, forced the owners to give more attention to the problems of marketing.

Any of the big mills could produce 500,000 board feet of finished lumber a week, and unless the manufacturer moved his stock promptly and regularly, he would find his yards, sheds, and storage spaces overflowing with lumber and his mill forced to shut down. Moreover, without constant and regular sales he would be without funds to continue his operations and would find both his capital and his workers' labor tied up in a large, static inventory.

There were several methods by which the manufacturer might handle the marketing of yellow pine. Some millowners sold their entire output to wholesalers or commission men. Others maintained sales organizations complete with their own traveling salesmen. They supplied thousands of independent retail yards throughout the country. Some of the largest producers owned chains of retail yards that marketed their products. At various times groups of mills attempted to organize sales companies to handle their total output. Fear of violating the provisions of the antitrust acts, however, usually prevented them from utilizing this method effectively.[3]

Wholesale dealers could be classed into two general groups. Dealers in the first group maintained yards in which they concentrated the cut of a number of mills, usually smaller operations, and provided equipment to dry, resaw,

[1] See chapter 2 for accounts of early mills and yards.
[2] Mary Austin Holley, *Texas*, pp. 71, 120.

[3] James Boyd, "Fifty Years in the Southern Pine Industry," *Southern Lumberman* 144 (December 15, 1931): 59–67.

Stacked lumber ready for market. *Courtesy of Ste-* *lections, Thompson photo albums.*
phen F. Austin State University, Forest History Col-

Yard scene showing poles and crossties ready *University, Forest History Collections, Thompson*
for shipment. *Courtesy of Stephen F. Austin State* *photo albums.*

and plane the lumber as required. Such dealers were prepared to sell dressed, dry lumber to retailers or large construction projects on demand. Some wholesalers in or near large cities maintained yards that could resupply local retailers promptly in whatever lengths and quantities were needed. The yard wholesaler, who bought from the mill or concentration yard in carload lots of lumber in specific sizes and grades and then sold to the retail yard an assortment of various items, performed a valuable service and provided an important link between the producer and the ultimate buyer. Wholesalers in the second group, "office" or direct wholesalers, maintained no yards at all. Such a wholesaler was a trader, a speculator who parleyed his knowledge of the lumber trade and current market conditions into a profitable business. He took orders for a fixed quantity of lumber from a factory or retail yard and then placed these orders with mills for direct shipment to the consumer. Or he might purchase a quantity of lumber from a mill and resell it at a profit. Such a wholesaler usually carried on his business in some metropolitan center. If he operated on a large scale, he employed a number of salesmen and buyers and perhaps maintained a traffic department to expedite shipments.[4]

At various times most of the East Texas lumbermen sold their products through wholesalers. The natural concentration centers of Texas pine were Saint Louis, Kansas City, and Chicago, and a steady stream of lumber-laden cars went north to these points. Secondary concentration points were Houston and Dallas, and important wholesalers operated from them. John Henry Kirby, with his usual propensity for multiplying corporations (see chapter 7), organized a wholesale outlet in Houston that dealt both in Kirby mill products and in the general market. Some lumber manufacturers were so impecunious at times that they would load a car with lumber and start it in the direction of the market, trusting in wholesalers with whom they had connections to dispose of it before it reached its destination. Some wholesalers specialized in these "transit-car" deals, and if they lacked a concentration yard, they frantically tried to sell the carload to some customer. This practice, of course, had a demoralizing effect on the general market and led to reckless price cutting. In the early days of their operations Kurth and Henderson on occasion started a carload of lumber toward Saint Louis expecting to sell it before its arrival. Apparently in most instances they were successful.[5]

Many lumber manufacturers employed commission agents and salesmen to market their products. The true commission man operated on a contract basis and had no responsibility for shipment, delivery, or payment. His commission was a set percentage of the order and often was paid one half on the acceptance of the order and the other half on final delivery and payment.[6] Many agents worked on a combination basis—part commission and part salary plus an expense account. For many years the Angelina County Lumber Company employed salesmen who traveled, saw prospects, and kept contact with the home office in Keltys. They recorded their expenses, item by item, in handy books provided for the purpose that were filed with the company for payment. The company sold extensively to retail lumberyards in the Middle West, evidently bypassing the wholesaler most of the time. The company kept a file index of several thousand retail yards, sash-and-door manufacturers, building-supply companies, and the like, in Michigan, Ohio, Indiana, Iowa, Illinois, and Wisconsin.[7] To the manager of a retail yard the visit of the lumber representative was often a welcome

[4] Stanley F. Horn, *This Fascinating Lumber Business*, pp. 184–89.

[5] Ibid., p. 188. For early operations of the Angelina County Lumber Company see Letterbooks, 1888–96, ACLCo. Papers, Forest History Collections, Stephen F. Austin State University Library, Nacogdoches; hereafter cited as SFA Library.

[6] Nelson Courtlandt Brown, *Lumber: Manufacture, Conditioning, Grading, Distribution, and Use*, p. 269.

[7] See account books and retail yard indexes, ACLCo. Papers. Forest History Collections, SFA Library.

break in the routine. He brought news of the outside world, the latest gossip of the trade, and often greetings from an acquaintance up the road. According to one such small-town yard manager, agents were a romantic lot:

Traveling salesmen I have known form an interesting chapter of my life. All will agree, I think, that lumber salesmen, as a class, are of the highest order. These Knights of the Road were always welcome. They brought a message, as it were, from the outside world, keeping us posted on various happenings of our fellow lumbermen. Withal, they were an honest, jovial lot and would not knowingly misrepresent their products in order to make a sale.[8]

Other companies employed no traveling salesman but sold their products through their own sales offices by means of letters and advertisements. The Thompson lumber enterprises maintained a central office in Houston from which the sales organization, headed by J. Lewis Thompson and a younger brother, Liggett, marketed about 80 million feet of lumber a year through both wholesale and retail channels. In addition to sales in about twelve middle-western states and Mexico, the Thompson office sold considerable lumber to James A. Thompson (the oldest of the sons), who ran a chain of yards in central Texas, on the Missouri, Kansas and Texas and International and Great Northern railroads.[9] In slow times, especially during periods like the Panic of 1907, Thompson would spice letters to customers with a bit of humorous doggerel, urging them to part with their money and buy now. One such catchy rhyme ran as follows:

Turn loose those orders.
Hurry, don't kick.
Out with your pencil;
Mark them all quick
Prices are steady,
Soon they will rise;

John Martin Thompson, founder of the Thompson lumber interests, ca. 1907. *Courtesy of Stephen F. Austin State University, Forest History Collections.*

Others are buying—
Nerve? Yes, but wise.

Timbers and yard stocks,
Uppers, kiln dried
Ceiling and flooring
Kept nicely tied.
Everything first class,
Rough Stock "tip top."

Lumber well-seasoned
Unknown to show rot.
Much could be written
Bold facts, with ease;
Ever need pickets?
Remember us, please.

Come
On![10]

Most finished lumber reached the ultimate consumer through the independent retail yard.

---

[8] Claude W. Bryant, *Lumbering Along in Texas*, p. 25.

[9] *American Lumberman*, September 26, 1908, pp. 126–33. A. C. Ford, a partner in the Grayburg mill, also operated a line of retail yards in southwest Texas and a wholesale business in Fort Worth, which provided an additional outlet for Thompson lumber.

[10] *American Lumberman*, September 26, 1908, p. 127. Note that the first letter of each line spells out Thompson Tucker Lumber Co.

The company office, ca. 1910. *Courtesy of Stephen F. Austin State University, Forest History Collections, Thompson photo albums.*

Company office, with secretaries. *Courtesy of Stephen F. Austin State University, Forest History Collections, Thompson photo albums.*

Here at last was the point of contact between the industry and the general public. The manager of such a yard was generally its owner, and he was an active, responsible, more or less permanent member of the community. Whether the purchaser was a householder who wanted to build a fence or a flower box or a contractor wanting lumber to erect a building, he dealt with the retail lumberyard. In the early days the retail yard sold only lumber; if the customer wanted other products, he went elsewhere. By World War I, however, most lumber dealers had become building-supplies dealers and stocked various materials and products in addition to lumber.[11]

The retail lumber business was (and is) heavily overcrowded and highly competitive. Every local lumber company competed with every other company. Dealers in hardwoods competed with dealers in softwoods. Sellers of southern pine competed with West Coast lumber dealers. Lumber competed against lumber substitutes. In each of these situations it was the retail dealer who provided the key, which usually was service. The dealer had to be

knowledgeable about materials and able to satisfy the customer that he was providing the best products available for the job. His inventory must be complete so that he had the desired lumber in stock. His deliveries must be prompt, and he must show a continuing interest in his customers' projects and problems. In all transactions the retail dealer sought to generate goodwill and gain the confidence of the general public. As one writer aptly said, "He has all his eggs in one basket," and he worked to gain and keep customers in his hometown and the surrounding territory.[12]

A representative retail lumberman was Claude W. Bryant. After serving an apprenticeship as an office clerk and salesman in several West Texas towns—Lott, Comanche, and Stamford—Bryant managed a yard at Sulphur Springs for the Harris-Lipsitz interests. In the 1920s he opened his own yard in Sweetwater, which he operated until the Depression. As an independent retail yard owner, Bryant was deeply involved in the affairs of the town. He was a staunch Presbyterian churchman, a Ro-

[11] Horn, *This Fascinating Lumber Business,* pp. 190–91; Brown, *Lumber,* p. 270.

[12] Ralph C. Bryant, *Lumber: Its Manufacturing and Distribution,* p. 387; Horn, *This Fascinating Lumber Business,* pp. 191–92. Brown, *Lumber,* pp. 271–72.

Stacked lumber in the open yard. *Courtesy of Stephen F. Austin State University, Forest History Collections, Thompson photo albums.*

tarian, a city councilman, and mayor for a time, and he was a director of the Lumbermen's Association of Texas. Friendly, outgoing, and hardworking, Bryant got on well with his customers and built up a business whose annual gross sales were over $150,000. His philosophy of business was simple: "Good stocks, good lumber, good service, and honest dealing."[13]

In reading Bryant's reminiscences, the reader can trace some of the economic and social history of the region. During the early years of Bryant's career in West Texas, around the turn of the century, the retail yard handled such items as cement in 400-pound barrels, lime in 200-pound barrels, cypress cisterns, cypress shingles, and wooden gutters. A regular feature of the early lumberyard was a supply of spindles, balusters, fancy railings, and gable brackets, all of which were in demand for the ornate Victorian houses of the period. After World War I the fashion in house building changed to less elaborate and more functional designs, and the lumberyard's inventory of these items disappeared.[14] In addition to serving business and residential building needs in the town, Bryant's lumberyards also sold large orders of lumber and related items to farmers and ranchers in outlying areas of the county. Usually the ranchers hauled the lumber in their own wagons, for they lived far from the railroad, and trucks did not appear in rural West Texas until the late 1920s.[15]

The effects of the Great Depression on the lumber industry—marketing as well as man-

[13] Bryant, *Lumbering Along in Texas.*

[14] Ibid., pp. 24, 30.
[15] Ibid., p. 16.

ufacturing—were well illustrated by the fate of Bryant's lumber company in Sweetwater. Lacking large credit reserves, Bryant was unable to cope with the rapid drop in construction of all kinds after 1930. Business fell off badly, accounts due became overdue, debts piled up, and finally, in 1932, Bryant was forced to sell the business at a great personal loss. As he ruefully commented, "No one lost anything on C. W. Bryant Lumber Co., but C. W. himself didn't fare so well. I wound up with just about what babies possess when they are starting out in the world."[16]

Another example of the family-operated retail lumberyard was the Hawn Lumber Company, of Athens, Texas. Founded by C. H. Hawn in 1881, the business grew and developed with the town. During boom years Hawn also operated several small sawmills and shipped finished lumber products to outlying farmers by wagons pulled by four-horse teams. The founder established a reputation for fair dealing, good stocks, and continuing interest in the community and his customers. The second- and third-generation Hawns continued the business along the same pattern and enjoyed a stable and prosperous business. In 1981 the Hawn Lumber Company had been in business without a break for one hundred years, owned and operated by the same family.[17]

Some lumbermen began with a single retail yard and later expanded to a string of yards plus a wholesale business and perhaps a sawmill. A few, starting with a single retail yard, emerged eventually as great lumber tycoons having important holdings in all phases of the

William C. Cameron. *Courtesy of Stephen F. Austin State University, Forest History Collections.*

industry. Among these were William Cameron; Robert A. Long, of Long-Bell Lumber Company; and John M. Foster. From slender beginnings these energetic entrepreneurs developed great industrial complexes of which their Texas holdings were only part. The retail yards of such corporations occupy a large place in the marketing picture of the Texas lumber industry.

Easily the dean of retail yards in the Lone Star State was William Cameron, of Waco. Born in Perthshire, Scotland, Cameron immigrated to the United States in 1852 at the age of eighteen. Like his fellow countryman Andrew Carnegie, the canny Scots lad was industrious, frugal, and ambitious. After serving in the Civil War, Cameron contracted to supply the Missouri, Kansas and Texas Railroad with ties and construction timbers. In 1867 he established his first retail lumberyard in Warrensburg, Missouri, and continued to place yards along the railroad line as the MKT built south.

[16]Ibid., p. 123. Claude Bryant's career as a lumberman did not end with this reverse. Thanks to Ernest Kurth and others, he became manager of the Clay Lumber Company, in Stephenville, before the end of 1932. This company, after reorganization, survived the Depression and expanded to include several additional yards in West Central and West Texas towns. He continued in this position for twenty-six years, until his retirement. As a career retail lumberman, Bryant was deeply interested in lumber and the lumber industry. "If you ever get pine splinters in your hands," he remarked, "you never get them out."

[17]"That Wonderful Year," *Athens* (Tex.) *Review*, August, 1962 (film tape of progress report, 1881–1946).

Cameron retail store, south Fort Worth, 1915.
Courtesy of Gulf Coast Lumberman.

In 1876, Cameron opened a yard at Waco and soon afterward established his headquarters there. He continued to expand his business, and by 1890 the William Cameron Company had more than sixty retail lumberyards.[18]

Perhaps because he was suspicious of the monopolistic tendencies of his competitors, Cameron branched out to acquire timberlands, six sawmills (including two at Saron and Carmona on the "Orphan Katy"), and various lumber-manufacturing and specialty shops. At Waco he established a sash-and-door company that eventually became (under his son) the largest standard millwork plant in the South. He also operated woolen and flour mills in a number of southern states, including Texas.[19]

[18] Walter Prescott Webb, ed., *The Handbook of Texas,* 1: 257–76; R. J. Tolson, *A History of William Cameron and Co., Inc.* Line yards were chains of retail yards owned by a single interest, often spaced along a given railroad.

[19] Tolson, *A History of William Cameron and Co., Inc.*; *Waco Tribune-Herald,* April 26, 1964; *Texas Almanac, 1958–59,* pp. 197, 318.

William Cameron was a human dynamo from whom energy seemed to radiate. He had a genius for work, not only mental and executive work but hard, physical, bodily toil, which brought "sweat to the brow" and left one weary at night. He was not above stepping in and working shoulder to shoulder with his laborers to show them just how a task should be done. He was impatient with indecision. He wanted his managers and foremen to have their own opinions and express them. Although he might disagree and argue heatedly with them, he respected them more for standing up for their views. He would not tolerate yes men. He hated details. Those he left to his partners and associates. He admired initiative and seldom told a manager how to do a job—just the results he expected. In his business ventures if a yard failed to show a profit, he quickly closed it out or sold it. He was constantly buying and selling lumberyards. He did not have time for a "losing proposition." He had little use for caste. He never judged a man by his family background or his clothes; instead, he weighed his performance. His three tests were "work, honesty, brains." Yet, according to his associates, with these driving characteristics there was kindness, a gentleness and charity that mellowed his character and made him an inspiration to his workers and partners. One employee expressed his feelings: "To have worked for him, or with him, was an enviable privilege, exceeded only by the inspiration which his example and success in life affords." [20]

A devoted family man, Cameron married twice and reared a family in Waco. He also kept in touch with his mother and other relatives in Scotland, sent them large sums of money, and made periodic trips back to the land of his birth. Writing to a sister in 1899, just a few weeks before his death from a stroke, he summarized his career:

"Now as to my Business career—it has been a prosperous one—now don't think me egotistical—I do this only when talking to a beloved Sister—I really don't know what I am worth—put it at $3,000,000, in your money $750,000 pounds sterling. Am the largest Manufacturer in the South—Lumber, flour—and Woolen Goods—I have 8 saw mills in the States of Texas and Louisiana employing about 3,000 men and upwards of 200,000 acres of Timber land manufacturing nearly one million feet per day—all of which I attend in person—Flouring mills take 10,000 bushels wheat per day—1740 pounds sterling per day to run them—The Woolen Mills employ some 800 men and women. My business points are some 100 miles apart. All are carried on by myself—son and two son-in laws [*sic*]. I can safely say that my business career is second to none—My credit is unlimited—The name of Wm. Cameron is known all over these United States.

His son, William Waldo Cameron, became president of William Cameron and Company after his father's death and continued to develop the business. Not yet twenty-one when he succeeded to the vast Cameron enterprises,

William Waldo Cameron surprised many associates by developing into a capable manager. By nature more conservative and cautious than his father, he built the company carefully and avoided speculative ventures. Like his father, however, the younger Cameron demonstrated ability to make decisions, delegate authority, and reward merit. Like his father he had little time for excuses and none for ostentation. By the time of the Great Depression of the 1930s the company business included more than eighty retail yards in Texas, Oklahoma, and New Mexico and the sash-and-door manufacturing and wholesale industry, which had nine branches and warehouses. Also like his father the younger Cameron was a philanthropist, and the city of Waco received Cameron Park as a memorial to the senior Cameron. [21]

One of the very large corporations that played an important role in Texas lumbering was Long-Bell. Like the William Cameron Company, it started very modestly as a retail yard. Robert Alexander Long was born in Kentucky in 1850 and moved west to Missouri as a young man. In 1875 he joined with Victor B. Bell to establish a retail lumberyard in Columbus, in southeastern Kansas. Soon they were adding other yards, and in 1884 the partners incorporated as the Long-Bell Lumber Company with headquarters in Kansas City. As Robert Long explained, "We grew yard by yard." Later manufacturing was added, and the company grew yard by yard and mill by mill. [22]

By 1902, Long-Bell had expanded to include 35 retail yards and had begun to build or buy sawmills and acquire timberland. By 1918 the company owned 118 retail yards, a dozen or so

---

[20] *Waco Tribune-Herald*, April 26, 1964.

[21] Ibid.; see also Tolson, *A History of William Cameron and Co., Inc.*, pp. 33–37; Webb, ed., *The Handbook of Texas*, 1:276. Many well-known lumbermen were at some time associated with William Cameron and Company, including E. H. Lingo, W. B. Brazelton, J. S. Mayfield, C. L. Johnson, H. A. Smith, G. M. Bowie, Willard Burton, and G. H. Zimmerman. These and many others would say that they served their apprenticeship under William Cameron.

[22] *American Lumberman*, March 30, 1918, p. 1; *Log of Long-Bell*, March, 1919.

Waco Sash and Door Company millwork plant.
Courtesy of Gulf Coast Lumberman.

sawmills, and more than 600,000 acres of timber. With mills in Arkansas, Louisiana, and Texas, as well as the Pacific Northwest, Long-Bell claimed to be the largest manufacturer and distributor of lumber in the United States.[23]

Long-Bell's Texas operations centered in the Lufkin Land and Lumber Company and the Fidelity Lumber Company (acquired from the Thompson family in 1912) at Doucette. Both of these were modern lumber-manufacturing plants, with rated capacities of 75,000 to 100,000 board feet a day. After acquiring the Doucette mill, Long-Bell further expanded and modernized it. As could be expected, the company's Texas timber holdings centered in Angelina, San Augustine, Tyler, and Trinity counties to support their mill operations. Most of the lumber manufactured in the Texas Long-Bell mills went to Kansas City and the company's retail yards in Missouri, Kansas, and Oklahoma.[24]

Robert A. Long was something of an enigma. In his black suit, high collar, and somber tie and with stern, unblinking eyes behind rimless gold spectacles, he frequently gave persons a strange, uneasy feeling. He was religious, conservative, and rather sanctimonious. He prided himself on his mastery of details and his knowledge of operations. At conventions and other gatherings he enjoyed astounding groups with his phenomenal memory. According to one biographer, he was a total abstainer: he neither drank nor smoked; he did not hunt, fish, or play cards, golf, or any other game; and he seldom attended the theater. He was a hard worker, a driving executive who regularly received and perused more than two thousand pages of reports a month concerning the operations of the company.[25]

Long took pride in "the family of Long-Bell." He had the company salesmen regularly attend schools, visit the mills, and assemble at the home office for briefings. He harangued his listeners on the difference between "small" and "trivial" matters and regularly stressed the high ideals of Long-Bell. The sales organization was huge. In 1915 the company had thirty-nine sales representatives scattered from Kansas City to Philadelphia and from Chicago to Houston. Long habitually called the salesmen "business associates" and maintained a punctiliously correct attitude in all personnel relationships. At the end of the 1920s, Long-

[23] *Log of Long-Bell*, June, 1919; *American Lumberman*, March 30, 1918, p. 55.

[24] K. G. Hanson (Long-Bell) to W. D. Oliver, June 5, 1958, Miscellaneous Papers, Forest History Collections, SFA Library.

[25] *American Lumberman*, March 30, 1918, p. 1. Long was also an early efficiency expert. Denouncing the high accident rate in the mills as wasteful, he promoted a safety-first movement to remove dangerous areas and operations. He considered whether a man was tall or short, strong, quick, and agile in placing him with the company. *Log of Long-Bell*, July, 1919.

Bell had exhausted the timber supply in East Texas and had closed its mills, having transferred its interests to the Pacific Northwest, where it built Longview, Washington, as a new center for its lumbering operations.[26]

A third major retail-yard chain that was Kansas City–based was the Foster Lumber Company. John McCullough Foster moved west from Pittsburgh shortly before the Civil War. He hauled freight, worked as a carpenter, and farmed near Leavenworth, Kansas, before he established his first retail lumberyard at Randolph in 1879. Taking his sons into partnership with him, Foster established additional yards in Kansas and headquarters in Kansas City. The sons took part in the Oklahoma land opening of 1893, and the family opened new lumberyards in that territory. By 1900, at the time of the death of John McCullough Foster, the company listed twenty-four retail lumberyards, mostly in Kansas and Oklahoma. Upon his death his son Benjamin B. Foster became head of the company.[27]

Much of the Foster story in Texas concerns timberlands and sawmill operations to support the string of line yards. As early as 1894 the company acquired 1,476 acres at two dollars an acre in Montgomery County near the present town of Fostoria (formerly Clinesburg). Two years later the Foster sons established the Trinity River Lumber Company as a wholesale house at Conroe but in 1897 moved to Houston. The Houston office marketed stock wholesale in Texas and Mexico and supplied the company's yards and wholesale business in Kansas City.[28]

Benjamin Foster invested in the Thompson and Tucker Lumber Company in 1902 and at one time held 44 percent of that company's stock. Then, in 1905, Foster began construction of a big new mill at Fostoria, which began operations the next year. The mill included two large double-cutting band saws with a capacity of 150,000 board feet a ten-hour day. At the time of its construction the mill was acclaimed as "the fastest mill in the South." Fostoria, situated on a branch of the Santa Fe Railroad, was cleaner and better equipped than the average Texas milltown. Its buildings were regularly painted, and it presented a pleasing, prosperous appearance. With a new mill, an attractive company town, and higher-than-average wages, regularly paid, the Fosters enjoyed good relations with their employees and had less than the usual turnover of workers.[29]

Despite the construction of the big mill at Fostoria, the Foster Lumber Company remained oriented to the wholesale and retail lumber business with its center of interest in the Middle West and the plains states. Texas supplied the raw materials and part of the finished lumber, which was sold through the central offices in Kansas City and Houston and the chain of more than sixty line yards stretching from Kansas to Wyoming. The company continued to add to its holdings of Texas timberlands until it held more than 130,000 acres in Montgomery, Polk, Harris, San Jacinto, and adjoining counties.

In Houston, native Texan Innis R. Palmer began as a bookkeeper for the Foster Lumber Company in 1909 and worked his way up until

---

[26] *American Lumberman*, January 20, 1906, p. 38; October 23, 1915. Despite his moral and religious attitude, some associates deeply resented Long's ruthless business drive and considered that he was not above sharp practices. Conversations with Hoxie H. Thompson, 1958; conversation with Mrs. Hoxie Thompson, 1962, Oral History Collections, Forest History Collections, SFA Library. Unless otherwise noted, interviews and conversations cited are in these collections, and all interviews were conducted by principal author Maxwell.

[27] Foster Lumber Company, "Through the Years with Fosters," Foster Lumber Company Papers, Forest History Archives, SFA Library.

[28] Ibid. In 1912, Benjamin Foster issued a three-volume work,

Standardized Plans for Line-Yard Building. These volumes gave approved plans for the construction of company line yards, ranging from large, urban operations to small, rural yards. Included also were specifications and estimates for their construction. As a consequence, Foster yards invariably presented a neat, efficient, uniform appearance.

[29] Ben Foster to Innis R. Palmer, May 25, 1940, Foster Lumber Company Papers. The company acquired and operated a number of sawmills in Texas towns including Leonidas, Clawson, Humble, Moscow, New Caney, New Waverly, and Keenan. Most of the mills were small and were closed after the completion of the mill at Fostoria. Foster Lumber Company, "Through the Years with Fosters," Foster Lumber Company Papers.

he became manager of the Houston office. At the mill at Fostoria, two generations of Dunnams, "Fox" and his son, Norman, served as foremen and mill managers. Although largely absentee, the Fosters were enterprising and considerate employers. Their business continued to prosper until and again after the Great Depression.[30]

Most Texas retail lumbermen and some manufacturers and wholesalers belonged to the Lumbermen's Association of Texas. This organization dated back to 1886, when a group of lumber manufacturers and dealers met in Austin with the express purpose of forming such an association. The sponsoring group included William Cameron; Ed Steves, a leading lumber dealer of San Antonio; O. T. Lyon, of Sherman; J. Witselle, of Corsicana; and Adam Van Patten, who was elected the first president. The association was initially called the Lumber Dealers' Association of Texas, but three years later the members changed the name to the Lumbermen's Association of Texas (LAT). The meeting place of the convention rotated, chiefly among the larger cities, Houston, Galveston, San Antonio, Waco, Fort Worth, and Dallas, and the members combined serious discussion with high jinks and recreation. In such an organization the position of secretary was often of crucial importance. In that position the LAT enjoyed the services of Carl F. Drake, of Austin, from 1886 to 1907, and of Sam T. Swinford, of Houston, from 1907 to 1913. Jack Dionne, editor of the *Gulf Coast Lumberman*, became secretary in 1918 and served until 1930 (it is of note that Dionne also served as secretary of the Texas Forestry Association during part of the same period).[31]

Many of the problems discussed at LAT conventions were very serious ones, and some threatened to break up the organization. There

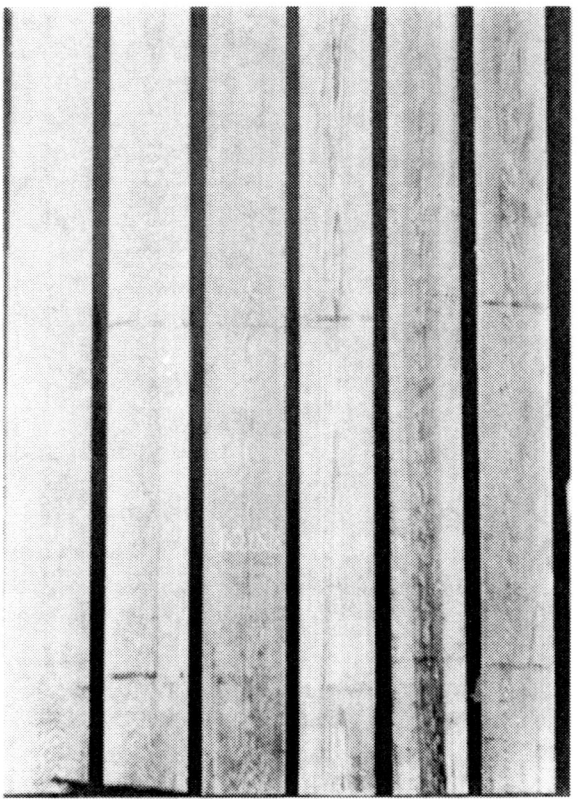

Examples of clear pine lumber, B & Btr. *Courtesy of Stephen F. Austin State University, Forest History Collections, Thompson photo albums.*

were chronic complaints against members who were both manufacturers and retailers. It was charged that they undersold independent retailers and that they intimidated the retailers by threatening to build more line yards unless independent merchants patronized their mills. The result was the formulation of a code of ethics that protected both manufacturers and independent retailers. The legal status of the association was also at times questioned. The attorney general of Texas investigated the organization to determine whether it was violating the antitrust laws. The association established the Lumbermen's Underwriters insurance institution and, in 1922, a traffic department. To help the members pursue their legal rights, George H. Zimmerman, of William Cameron and Company, compiled a "red book" summarizing and explaining the lien

[30] Foster Lumber Company, "Through the Years with Fosters," Foster Lumber Company Papers, interview of Hubert Smith with James Kersh.

[31] *Gulf Coast Lumberman*, 51, no. 8 (November, 1963): 36–37.

laws of Texas. By 1930 the Lumbermen's Association of Texas had passed through a series of trials and tribulations to emerge as a cooperative service organization. Its stress on modern merchandising methods and better building practices greatly aided both the retail dealers and their customers. As one writer said, "The History of the Lumbermen's Association of Texas is practically the history of the retail lumber business in that State."[32]

In the half century from 1880 to 1930, the bonanza period in the Texas lumber industry, the retail yards changed perhaps as much as did logging practices and technology in the sawmills. In the early days lumberyards were always built along the railroad tracks. An observer would not have found an office or a showroom there or any displays of building materials. Instead he would have found a "rathole" for an office, frequently little more than a shack, where cobwebs abounded and mouse nests clustered in the corners. The only place to sit was an empty nail keg, and the manager had his papers and invoices in a rolltop desk at which he would sit with his hat on to figure an estimate on the customers' posed purchases. Outside most of the lumber was stacked random fashion in the open air with inadequate supports. Only flooring and other top-grade lumber were protected by a shed. Once the customer had been given his estimate, he would probably walk down the track to the yard of a competitor, who would try to undercut the first bid.[33]

By the 1920s all this had changed. The retail lumberyards had moved uptown to more convenient locations. A more modern office with adequate furnishings and a display of the latest materials and fittings would invite the wife as well as the husband to help make selections for remodeling or building a house. The manager no longer waited for business to come to him but actively merchandised and advertised his building materials and sought to help customers select the materials best suited to their particular needs. In short, he saw his job as one to "translate building material into the language of homes."[34] In the yard proper most if not all of the lumber was stored in covered bins. The introduction of grade-marked, Southern Pine Association—inspected lumber in the 1920s revolutionized the merchandising of lumber products and greatly improved the confidence of builders in the quality of their purchases. Companies like the Foster Lumber Company led the way in making the lumberyard a convenient, modern, attractive, well-stocked shopping center for all sorts of building supplies. This revolution in retail lumber merchandising was characterized by the cartoon figures "Mr. Pip and Mr. Pep," which Jack Dionne created in the *Gulf Coast Lumberman* to illustrate the broad difference between the progressive dealer and the "what-was-good-enough-for-father" school.[35] This was marketing, and the retail lumber dealer provided the point of contact between the lumber industry and the buying public. The industry was often judged by his performance.

[32] Ibid.; Bryant, *Lumbering Along in Texas*, p. 102; Boyd, "Fifty Years in the Southern Pine Industry," pp. 59–67. In 1959 LAT built its own home and headquarters in Austin, largely donated through the cooperative efforts of members of the industry. Many manufacturers continued to hold memberships, and a few have been elected to office, but essentially the Lumbermen's Association of Texas has been the retail dealers' organization and their voice. Bryant, *Lumbering Along in Texas*, pp. 193–94; *American Lumberman*, April 12, 1902, p. 20.

[33] *Gulf Coast Lumberman*, November, 1963, p. 36.

[34] Ibid., p. 94.

[35] Ibid. The Lutcher and Moore Lumber Company pioneered in the grade marking of yellow-pine lumber in 1924. Its success led the Southern Pine Association to adopt grade-marked standards for the industry. See "The Story of Grade Marked Yellow Pine Lumber," Lutcher and Moore Lumber Company Papers, Forest History Collections, SFA Library.

CHAPTER 7

A Prince, a Baron, and a Lady

He doth bestride the narrow world
Like a Colossus . . .—William Shakespeare

THE lumber industry during the half century of the bonanza era in the Gulf Southwest was by no means a cold, impersonal operation. Individual men and women directed and controlled the decision-making process of even the largest companies and placed their personal stamp on the character of the enterprise. In the typical large company, whether corporation or partnership, only a few men—often only one—controlled the success or failure of the venture. They were a varied lot in background and included native sons and newcomers from the Old South, the Great Lakes states, the East, and even Canada and Europe. They had certain characteristics in common: they were bold, aggressive, optimistic, adventuresome risk takers. They were entrepreneurs in the classic definition of the term. As a group they were impatient with government restraints or restrictions, paternalistic toward their workers, but highly resistive to the demands of organized labor. They heeded the advice and warnings of conservationists only when they

became convinced that sound conservation practices were sound business investments for the future. In short, the lumber barons of the Gulf Southwest who flourished between 1880 and 1930 were not unlike their counterparts in the empires of steel, oil, meat processing, and finance elsewhere in the United States at the same time.

There were far too many lumber barons to attempt to sketch them all. Many established lumber dynasties whose names became household words in the Gulf Southwest and continued to the second and third generations. The three families presented in this account are representatives of those who participated in the bonanza. Their stories are in many respects similar, but each family had some unique aspect or twist that made it different from the others. Collectively, together with the persons discussed in other chapters of this book, they form a who's who of the great names of lumbering in the Piney Woods.

John Henry Kirby

AMONG Texas lumbermen none was more colorful or better known nationally than the flamboyant, empire-building native son John Henry Kirby. Like most of his contemporaries

Material in this chapter dealing with Mrs. Knox substantially appeared in *Texas Forestry* 14, no. 4 (April, 1974): 28–29, and is used with permission of the Texas Forestry Association.

in the industry, Kirby was a complex individual who did not fit easily into a stereotype. At the peak of his career he was described as large, ruddy-cheeked, and gregarious, with great personal charm and a persuasive manner. He was a friend of presidents, governors, senators, and other personages and at the same time prided himself on his personal friendships with his employees and their families. Kirby inspired warm devotion, developed close, lasting associations with his colleagues and simultaneously drew sharp, bitter criticism for some of his labor policies and his strong antiunion stance. As the largest lumber manufacturer in the Gulf Southwest, he was perhaps the archetype of the Texas lumber baron of the bonanza era.

The Kirby family was of pioneer stock who had followed the frontier from Virginia through North Carolina, Kentucky, and Mississippi to Texas before the Civil War. His father, John Thomas Kirby, settled on a farm in Tyler County, Texas, near Peach Tree Village, where John Henry was born on November 16, 1860.[1] Like perhaps most other boys growing up in the rural South during this period, young Kirby experienced poverty and unending grinding toil. Unlike his neighbors, however, John Henry exhibited at an early age a desire for education and knowledge, a characteristic perhaps acquired from his mother, for John Thomas Kirby reportedly had no time for schools or "book larnin'." By the time he was sixteen, John Henry had attended every school within a radius of six miles of his home, often working to pay for his keep. After a school year at the county seat, Woodville, young Kirby worked as a deputy sheriff for a time to save money to attend Southwestern University, at Georgetown. When his money ran out, he found a position, thanks to State Senator Bronson Cooper, as calendar clerk in the Texas legislature, a job he held for three years. During that time he read law in Cooper's office, and in 1885 he was admitted to the bar. He set up a partnership to practice law in Woodville, where his work mainly consisted of untangling land titles and buying timberlands for clients.

John Henry Kirby, ca. 1901. *Courtesy of Stephen F. Austin State University, Forest History Collections.*

As mentioned earlier, most East Texas farmers regarded the pine forests as of little value and a hindrance to the development of farming. The story was told that the elder Kirby bought his wife a sewing machine from a traveling salesman and in payment deeded him a tract of pineland. A year later the salesman returned and traded the land back for "a cow or yoke of oxen," something he could sell.[2] John Henry disagreed with this assessment of his region's chief resource. From an early age he had had an attachment for the towering longleaf trees, and he soon developed an appreciation of their economic worth. As he later explained:

I had no money and I knew but few people of influence. Yet I did realize what was going on in my state. I saw that the bounds of western settlement

[1] Kirby's ancestry was mostly English and Scottish. One of his grandmothers was Elizabeth Longino, who was of Italian descent. When Kirby's friends said that he looked like a Roman senator, he often joked about his Italian blood. "Kirby Memoranda," Cooper K. Ragan Papers, University of Houston Library; hereafter cited as Ragan Papers; Mary Lasswell, *John Henry Kirby: Prince of the Pines*, p. 3.

[2] "Kirby Memoranda," Ragan Papers; Jack Dionne, *A Brief Story of the Life of John Henry Kirby*; Lasswell, *John Henry Kirby*, pp. 18–35, 44.

were being pushed out upon the barren prairies where there were no building materials. I knew that those settlers must look to the pine lands to supply their growing needs. It was then easy for me to understand the value of trees and I set out with the purpose of acquiring some of them as rapidly as I could.[3]

While practicing law in Woodville, Kirby made the acquaintance of a group of Boston capitalists that included Nathaniel D. and George Z. Silsbee. They had a large tract of land on which they were attempting without much success to run a ranching operation. Kirby became their Texas representative and lawyer and soon convinced the Boston group that they would make more money growing trees and buying timberland. As a result the Boston investors and Kirby formed land companies in Texas and Louisiana that soon acquired more than 250,000 acres of virgin pinelands containing an estimated three billion board feet of timber. In 1893 Kirby began building a railroad (the Gulf, Beaumont and Kansas City) to open up these holdings for potential lumbering operations, and by 1897 he had completed the road as far as Roganville, in Jasper County, and had plans to continue it to San Augustine, Center, and Marshall. In 1900 Kirby sold this road to the Atchison, Topeka and Santa Fe Railway, thus bringing that great transcontinental line into his circle of supporters. In addition, the Santa Fe lent money to Kirby for the purchase of more timberlands. This transaction began an intimate and friendly relationship between Kirby and the officials of the Santa Fe that continued throughout his life.[4]

Until 1900, Kirby had thought of himself not as a lumberman but rather as a lawyer and buyer of timberlands. True, he had acquired some small sawmills, but he had also built and run a railroad, though he was not a railroader. He had also engaged in real-estate development in Houston and had profited from the growth of that budding metropolis. He was chiefly known as an energetic man with enthusiasm to complete any task he undertook. Perhaps his greatest assets were his charm and personal magnetism, coupled with a warm, friendly smile. There was a popular story about an eastern banker who made inquiries of a Texas friend about Kirby's credit rating. The Texan replied, "I don't know a thing about him except he has a million-dollar smile." "You are right," said the easterner. "Young Kirby has just smiled me out of a million." At age forty John Henry Kirby was already wealthy, with a successful career behind him. His life as a great lumber tycoon was just beginning.[5]

The year 1901 was a good year for business enterprise and expansion. William McKinley was president, and the country was on an economic upswing after a short, successful war. That year John P. Morgan combined eight steel producers into the United States Steel Corporation, capitalized at more than $1 billion and described as the largest corporate structure in the country. The same year, as a result of the great Spindletop oil strike, new major oil companies, such as Texaco, Gulf, and Sun, emerged to challenge the monopoly of the Standard Oil Company. Kirby was keenly aware of these developments and the opportunity for similar consolidations in the lumber industry. Lumbering was growing rapidly in Texas, but was characterized by a great num-

---

[3] "He was the Plowboy of Peachtree Village—but He Became Prince of the Pines," *East Texas* 1 (December, 1926): 27–28.

[4] Rankin S. Nowlin, "Economic Development of the Kirby Lumber Company of Houston, Texas" (master's thesis, George Peabody College, Nashville, Tenn., 1930), p. 6; "Timber Resources of East Texas, Their Recognition and Development by John Henry Kirby through the Inception and Organization of the Kirby Lumber Company of Houston, Texas," *American Lumberman*, November 22, 1902; "Kirby Memoranda,"

Ragan Papers; Robert S. Maxwell, *Whistle in the Piney Woods: Paul Bremond and the Houston, East and West Texas Railway*, pp. 51–52; Keith L. Bryant, Jr., *History of the Atchison, Topeka and Santa Fe Railway*, p. 189.

[5] *Houston Post*, December 12, 1937; *Beaumont Enterprise*, November 6, 1955; "Kirby Memoranda," Ragan Papers; Lasswell, *John Henry Kirby*, pp. 49–50.

ber of relatively small companies of limited re-sources usually confined to a single millsite. He was certain that the time was ripe for the for-mation of a large, consolidated lumber corpo-ration capable of handling any volume of busi-ness from the eastern or overseas trade. He was also aware that much of the best longleaf-pine timber, including the holdings of the Boston group, lay only a relatively few miles north of Spindletop and might well include oil deposits in its subsurface sands.[6]

At the annual meeting of the American Lumbermen's Association held in March, 1901, in Chicago, Kirby pursued this theme in conversations with other lumbermen and banking associates. As a result of these talks Kirby began negotiations with Patrick Cal-houn, a New York attorney, and J. Wilcox Brown, a Baltimore banker. The group agreed to set up two corporations that would support each other in exploiting all the resources of the region. In July, 1901, Kirby and his associates announced the organization and chartering of the Houston Oil Company and the Kirby Lumber Company. The oil company, capital-ized at $30 million, would own and control more than a million acres of East Texas timber-lands and would engage in oil exploration and development. The lumber company, capital-ized at $10 million, was to own and operate fourteen to sixteen sawmill facilities and re-lated properties. The two ventures were tied together by a giant stumpage contract which called for the oil company to provide to Kirby no less than 350 million board feet of timber a year at prices ranging from $3 a thousand to $5 a thousand after the fourth year. Kirby fur-ther agreed to provide $2.5 million in working capital for the lumber company. Calhoun would provide working capital for the oil com-pany from the sale of stock, the marketing of

"timber certificates," and a $500,000 loan from Brown Brothers Bank of New York.[7]

The Houston press hailed Kirby as a finan-cial genius who had organized the two largest corporations in Texas. In November, 1901, his friends honored him at a lavish banquet that was attended by three to four hundred guests led by Texas Governor Joseph D. Sayers. The speeches on this occasion heaped fulsome praise on the boy from Peach Tree Village who had made good. Speakers described Kirby as "Houston's first citizen," the "father of indus-trial Texas," and "prince of the pines." Kirby in turn described the nature and organization of this financial venture; praised his associates, Calhoun and Brown; and promised to direct the affairs of the companies so as to promote the prosperity and progress of Texas. For a man of forty-one it was a heady experience.[8]

The enterprise, however, did not go well. Despite the number of lawyers involved, the papers listed the properties in only a general way and were vague about Kirby's and Cal-houn's obligations to their companies and to each other. Further, Kirby was unable to raise the $2.5 million in cash promised for operating capital. The sawmill properties, though listed at a capacity of 350 million board feet a year, were able to produce only about two-thirds that amount. Consequently, Calhoun was also short in his contribution to the common ven-ture. In addition Kirby had difficulty securing the transfer of about 179,000 acres of virgin longleaf timber owned by the Kountze broth-ers. Without this tract of valuable land all the estimates of worth and production figures would have to be considerably reduced.[9]

Soon both companies were in financial

---

[6] John O. King, *The Early History of the Houston Oil Com-pany of Texas, 1901–1908,* pp. 5, 12–13; James A. Clark and Michael T. Halbouty, *Spindletop,* pp. 79–81; Arthur M. John-son, "The Early Texas Oil Industry: Pipelines and the Birth of an Integrated Oil Industry, 1901–1911," *Journal of Southern His-tory* 32 (November, 1966): 516–28.

[7] Patrick Calhoun was a descendant of the South Carolina statesman John C. Calhoun and thus had automatic entree into postbellum society in the South. He was known as an aggressive, ruthless lawyer perhaps given to sharp practices to gain his ends. King, *The Houston Oil Company,* pp. 7–9, 15; "Kirby Memo-randa," Ragan Papers.

[8] *Houston Chronicle,* November 13, 1901; King, *The Hous-ton Oil Company,* pp. 1–5; Nowlin, "Economic Development of the Kirby Lumber Company."

[9] King, *The Houston Oil Company,* pp. 32–49.

straits, and the principals began denouncing each other for failure to provide funds as agreed upon and apparent bad faith. Each filed suits, which were met by countersuits, and by January, 1904, both companies were placed in receivership. The complicated litigation dragged on for four years, during which time various interested parties, such as the Boston syndicate, the AT&SF Railway, and the Brown Brothers Bank, intervened in the controversy. In time Kirby saw the legal maneuvers as an attempt by Calhoun and his eastern associates to "take" the country boy from Texas and squeeze him out of the valuable timber properties. Calhoun charged that Kirby had misrepresented the value of his properties and had defaulted in his obligations. Eventually the several claimants negotiated a compromise settlement ending the receivership. Kirby surrendered any claims to the Houston Oil Company properties, including title to about 800,000 acres of timberland. He gained confirmation of his ownership of the sawmills, yards, tramroads, and other properties. A somewhat reduced cutting contract was declared in force, and the stumpage rights to the oil-company lands continued to be the basis for the operations of the Kirby mills. Although he emerged from this prolonged controversy somewhat reduced in circumstances, Kirby's energy and resourcefulness had prevented Calhoun and the eastern bankers from, as he said, "plucking him like a chicken." The long bout in court apparently did not tarnish his Texas image, nor did it alienate his friends in the Boston syndicate or the Santa Fe Railway.[10]

Despite the receivership Kirby steadily organized and upgraded his lumber operations during these years. Neither the number nor the locations of the mills remained static, for he consolidated some mills, closed others because of worn-out or inefficient machinery, lost some to fires, and built new, more modern mills,

such as Browndell, in northern Jasper County, and Bessmay, named for his daughter, in southern Jasper County. A representative report of Kirby's Texas Mills was that of 1907, supplied by the *American Lumberman*:

Kirby Lumber Company (Houston) Production for 1907

Beaumont	17,059,000	board feet
Bessmay	57,724,000	
Browndell	36,033,000	
Call	40,932,000	
Ford's Bluff	16,804,000	
Fuqua	15,983,000	
Kirbyville	43,366,000	
Mobile	14,277,000	
Roganville	14,488,000	
Silsbee	12,102,000	
Village Mills	16,992,000	
Woodville	15,348,000	
Total	301,108,000 board feet[11]	

Although no single mill approached the output of some of the largest Texas producers, Kirby's total made the company easily the largest manufacturer of lumber in the state, perhaps in the entire South. No other Texas company reported a cut of as much as 100 million feet. As Kirby had promised when he organized the company in 1901, he had put together a corporation that could handle any contract, regardless of size or requirements.[12]

To support his sawmills, Kirby organized thirteen or fourteen logging camps in the same

---

[10] Cooper K. Ragan to John O. King, August 18, 1958, Ragan Papers; Sam Hanna Acheson, *Joe Bailey: The Last Democrat*, pp. 176–77; King, *The Houston Oil Company*, pp. 72–87; Lasswell, *John Henry Kirby*, pp. 96–114.

[11] *American Lumberman*, April 4, 1908, p. 65. When it was new in 1895, the Village Mills sawmill claimed a world's record for lumber cut in a single day's run. The feat was celebrated in news reports and poems. Lasswell, *John Henry Kirby*, pp. 65–66.

[12] *American Lumberman*, April 4, 1908, pp. 65–66; Nowlin, "Economic Development of the Kirby Lumber Company."

areas. Under Kirby's key principle of centralization and specialization, the mill managers did not have the responsibility for procurement; they were responsible only for production. Logging bosses superintended the felling of trees, cutting the trunks into logs of the desired length, and transporting them to the mills.[13]

As much as possible Kirby also standardized his logging methods. Basically logging for the Kirby Lumber Company was a rail-and-tramroad operation. From connections with the Santa Fe or Texas and New Orleans main lines, tramroads extended into the forests, and from them branches ran like tentacles into every section of the stand. After the fallers and buckers had done their work, steam skidders equipped with power drums and long cables brought the logs to trackside. There the steam loader, equipped with a crane, picked up the logs and placed them on a log car. This method was both quicker and cheaper than using mules or oxen to drag out the logs, but it was harmful to the young growth of the forest and left much of the cutover region with a rather "desolate appearance."[14]

Kirby also took steps to consolidate such mill services as commissary and medical care. He established a central commissary warehouse at Beaumont (directed for years by M. J. Godfrey), from which goods went out to the local commissaries on the T&NO or AT&SF. If certain items did not move at one town, he shipped them up the line to another store, urging the manager to make a special effort to dispose of them. Except in Beaumont most business at Kirby company stores was done with merchandise checks.

In like manner he established a hospital at Beaumont with several doctors in residence. He sought to have a doctor at every mill town, and if a doctor was not available, Kirby supplied a "medic." To pay for medical services, he set up a hospital plan under which he deducted one dollar a month from the wages of each of some five thousand employees. Because most of his mills and logging camps were directly connected with Beaumont by railroad, any seriously injured or ill worker could be transported relatively quickly to the Kirby hospital.[15]

The headquarters of this lumber empire was in Houston, where Kirby eventually built the Kirby Building to house his many and varied affairs. There Kirby and his lieutenants directed promotional and advertising campaigns for Kirby lumber, managed transportation problems with a score of rail lines, developed an extensive export trade, supervised the myriad purchases necessary for such an empire, and constantly sought opportunities to expand his business horizons.[16]

Public officials and the Houston press lionized John Henry Kirby. He was king of the Houston Cotton Carnival in 1900, and a few years later his daughter reigned as queen. He also served as president of the Texas exhibit at the Louisiana Purchase Exposition in Saint Louis in 1904. He was appointed a regent for the University of Texas in 1911. A staunch "Bourbon Democrat," Kirby regularly attended the Texas state Democratic conventions as a delegate from Houston and presided over the convention as chairman in 1904. The voters of his adopted city elected him to the state legislature in 1913 and again in 1927. The *Houston Post* described him as unique— "only one John Henry Kirby comes with a

[13] "Timber Resources of East Texas," *American Lumberman*, November 22, 1902, p. 27.

[14] *American Lumberman*, February 15, 1902; November 15, 1902; November 22, 1902. In 1902 Kirby accepted assistance from the U.S. Division of Forestry to examine his forest and advise him on a practical management program. As a result Kirby began developing his program of selective cutting and sustained-yield forestry. "The Kirby Story: 50th Anniversary of the Founding of a Lumber Empire," *Gulf Coast Lumberman*, July 15, 1951.

[15] *American Lumberman*, November 22, 1902, pp. 65–69.

[16] *American Lumberman*, November 22, 1902, p. 5; July 4, 1903, pp. 50–51; July 18, 1903, p. 41; February 6, 1904, pp. 22–23. In 1916 the Houston Oil Company transferred its surface acreage to the Southwestern Land and Development Company, thus completely separating the surface, the minerals, and the timber of the tract of land assembled by Kirby in 1901. King, *The Houston Oil Company*, pp. 88–89.

John Henry Kirby and President Warren G. Harding, 1922. *Courtesy of Stephen F. Austin State University, Forest History Collections.*

world." Many friends repeated the comment made by a black porter in Beaumont when asked who the distinguished gentleman was. "Cap, that's John Henry Kirby," said the porter, "and when he crows, it's daylight in East Texas."[17]

Kirby became almost as well known nationwide as he was in his home state. He was one of the founders of the Southern Pine Association and in 1922 served as its president. For four years, from 1917 to 1921, he was president of the National Lumber Manufacturers Association. During World War I he was lumber administrator for the South in the Emergency Fleet Corporation of the United States Shipping Board. After the war he served in President Warren G. Harding's Conference on Unemployment. During the 1920s the press reported Kirby at conferences from Seattle to New York and Washington, D.C., often as a spokesman for the lumber industry, sometimes

as a conservative Democrat, and always as a colorful Texan with a quotable comment.[18]

The Great Depression hit Kirby hard. Although he had millions in various properties, stocks, bonds, and other securities, many of these became unmarketable even at deflated prices. At the same time his liabilities, short-term borrowings, taxes, and endorsements pressed in upon him. In 1933, John Henry Kirby filed a voluntary personal petition of bankruptcy with the federal court in Houston. The Kirby Lumber Company was not involved in these proceedings and continued operations, though lumber prices had declined to below production costs. The Santa Fe Railway advanced loans to the company against capital stock and assumed control of the corporation. Eventually the Kirby Lumber Company became a wholly owned subsidiary of the Santa Fe. Kirby continued to enjoy the title of president of the Kirby Lumber Company and draw a salary until his death in 1940 at the age of eighty.[19]

It is difficult to evaluate such a man as John Henry Kirby. To his employees he was a combination of indulgent godfather and slave driver. He prided himself on knowing the names of most of the children in the company towns and on being a "pal" to his workers. No Christmas passed without Kirby handing out gifts totaling thousands of dollars. One year he distributed some twenty-five hundred Bibles; another year he passed out five-dollar gold pieces. He often played Santa Claus at the Christmas programs. Literally scores of young men and women, many of whom rose to prominent positions in business or the professions, thanked John Henry Kirby for funds that enabled them to go to college.[20]

To his production chiefs, mill managers, and

[17] *St. Louis Post Dispatch*, January 18, 1903; *Houston Post*, December 12, 1937; *Beaumont Enterprise*, November 6, 1955; Lasswell, *John Henry Kirby*, pp. 70–74.

[18] John M. Collier, *The First Fifty Years of the Southern Pine Association*, pp. 44–45, 171; James E. Fickle, *The New South and the "New Competition,"* pp. 95–103.

[19] "Kirby Memoranda," Ragan Papers; Lasswell, *John Henry Kirby*, pp. 186–200; Bryant, *History of the Atchison, Topeka and Santa Fe Railway*, pp. 189, 364.

[20] *American Lumberman*, December 30, 1916; Kirby Personal Papers, University of Houston Library; *Beaumont Enterprise*, November 10, 1940.

woods superintendents Kirby was a hard-riding driver who at times seemed almost insatiable. He proposed to pay managers in proportion to what they earned for the company and to provide a sliding scale in salaries to reflect productivity. A humorous story about Kirby's drive for efficiency and cost reduction has long since become part of the folklore of East Texas:

Kirby dreamed one night that he had died and gone to hell. When he awoke, he found himself in the grasp of two brawny black men who were carrying him off. He asked where they were taking him and they replied that they had orders to take him to the Lake of Brimstone and throw him in. Kirby at once retorted, "One of you lay off, and the other get a wheelbarrow. We don't need two men for this little job."[21]

All his life Kirby was an adamant foe of all efforts to unionize the sawmills of the South and referred to union organizers as "latter-day carpetbaggers." He continued to maintain a ten-hour day in all his lumber operations until late in the 1920s and only reluctantly reduced the hours of labor in his mills to nine in 1929. He never accepted the "principle" of the eight-hour day. United States Department of Labor investigators criticized his continued use of merchandise checks instead of cash on most paydays and characterized the workers' houses in his company towns as "drab, dingy, and unsightly." Throughout his career Kirby insisted on running his business according to his own concepts, without any interference from the state, the federal government, or the unions.[22]

Yet Kirby was not entirely unprogressive. He supported the conservation programs of the Texas Forest Service and was a friend of W. Goodrich Jones. He reformed his cutting practices and developed a sustained-yield rotation. Working with the Texas Forest Service, he began experiments with reforestation on his cutover lands. On a surprising number of Texas public boards and commissions one found Kirby serving as a member, often as chairman.[23]

For more than thirty years John Henry Kirby was the leading figure in the lumber industry of Texas and the Gulf Southwest. Displaying tremendous energy and drive, great personal magnetism, a confidence-inspiring presence, a venturesome spirit, and a thorough knowledge of pine timber, Kirby took advantage of a combination of fortunate circumstances to build a great lumber empire. At its height Kirby held hundreds of thousands of acres of the finest longleaf-pine stands in the entire South. Among the feudal baronies of East Texas during the bonanza era, Kirby held the largest and the most absolute fief. He was, indeed, "prince of the pines."

The Kurths and the Hendersons

IN marked contrast to John Henry Kirby, who was a native-born Texas farm boy, Joseph Hubert Kurth was an immigrant from Germany. According to one of his descendants, who searched out the records of his birthplace, Joseph Kurth (recorded as Simon Joseph Kurth in the local lists of births) was born in the Bonn area on July 3, 1857. He became a passionate Rhinelander and hated the Prussians who dominated the newly formed German Empire. In 1877, at nineteen, he refused to serve in the

---

[21] *American Lumberman*, July 18, 1903; conversation with Francis Ingraham Tucker, Nacogdoches, 1967, Oral History Collections, Forest History Collections, Stephen F. Austin State University Library, Nacogdoches; hereafter cited as SFA Library. Unless otherwise noted, interviews and conversations cited are in these collections, and all interviews were conducted by principal author Maxwell.

[22] George T. Morgan, Jr., "No Compromise—No Recognition: John Henry Kirby, the Southern Pine Operator's Association, and Unionism in the Piney Woods, 1906–1916," *Labor History* 10 (Spring, 1969): 193–204; Peter A. Speek, "Notes on Investigations of Three Texas Lumber Towns," in *Reports of the United States Commission on Industrial Relations, Department of Labor* (microfilm), SFA Library; Fickle, *The New South and the New Competition*, pp. 289–98; Lasswell, *John Henry Kirby*, pp. 186–200.

[23] "The Kirby Story: 50th Anniversary of the Founding of a Lumber Empire"; Robert S. Maxwell, "One Man's Legacy:

Joseph H. Kurth. *Courtesy of Stephen F. Austin State University, Forest History Collections.*

Simon W. Henderson. *Courtesy of Stephen F. Austin State University, Forest History Collections.*

imperial army and as an act of defiance threw stones into a window of the German headquarters in Bonn. The authorities tried to arrest him, but he fled and managed to escape from the Bonn region. Other relatives, somewhat more romantic, pictured him as a student in the University of Bonn who became involved in a student riot. According to them, the uprising began because of the arrest by German authorities of a popular liberal instructor, and Kurth joined others in stoning the burgomeister's house. He then fled, fearing arrest.[24]

Whatever the reason, Joseph Kurth soon de-

determined to leave Germany. After borrowing money from relatives, he stowed away on a ship bound for the Texas Gulf Coast. When he was discovered, he was put to work shoveling coal to pay for his passage. In this manner young Joseph Kurth arrived in Galveston, Texas, in 1878, with virtually no possessions except the clothes on his back. He was unskilled in any craft or business, and he spoke only a very broken, textbook English.[25]

Young Kurth soon went to work as a laborer at a Galveston sawmill. He moved from one job to another, performing various tasks and learning the duties of the job above his. He worked in a number of South and Central Texas towns, including Seguin, Willis, and Hartley. At Hartley he met and married Hattie

W. Goodrich Jones and Texas Conservation," *Southwestern Historical Quarterly* 77 (January, 1974): 376–77; conversations with John Kirby Herndon, vice-president, Kirby Lumber Company, Silsbee, Texas, 1958, 1960, author Maxwell's collection.

[24] Conversation with Dr. Robert L. Kurth, Nacogdoches, May 3, 1982; records and letters from Endenich, Germany (all in the possession of Dr. Kurth, Kurth Family Papers, Lufkin, Texas, hereafter cited as Kurth Papers). Conversation of Mrs. Robert L. Kurth with Evelyn Nichols, Lufkin, June 16, 1965, Kurth Papers. The records list Joseph Kurth's father as a

nailsmith who died in 1858, the year after Joseph was born. It is unlikely that the son of an artisan in those circumstances could have attended a German university in the 1870s.

[25] Conversation with Dr. Robert L. Kurth, May 3, 1982, Kurth Papers; *Lufkin Daily News*, June 16, 17, 1930; March 15, 1966.

Glenn, the sister of a fellow worker. The Kurths had six sons, five of whom lived to maturity and participated in various aspects of the family enterprises. In order they were Joseph H., Jr., Ernest L., Roy W., Melvin E., and Robert L. The family lived frugally, and although he was paid largely in merchandise checks, Kurth regularly saved enough to turn some back for cash. He soon bought a small sawmill in Polk County, south of Corrigan, on the Houston, East and West Texas Railway, at a point he called Kurth Station. There he learned about all phases of sawmill work, and there also he met Simon W. Henderson, a general merchant in Corrigan.

Simon Woods Henderson was born in 1859 on a farm in Georgia, one of fourteen children. His educational opportunities were meager and by his own estimate did not exceed a total of six months of formal classes. At an early age he went to work for himself, clearing land and doing rough carpentry work on small houses, for which he earned about 50 cents a day. In 1883 he migrated to Texas, settling in Corrigan, in Polk County, which had only the year before become the junction point for the Houston, East and West Texas Railway and the Trinity and Sabine Railway. Henderson went to work for a tie-cutting contractor and within a year purchased the business. In 1886 he acquired a general mercantile business in Corrigan and in a few years had amassed savings of about $15,000. Friendly and easygoing, Henderson was also an excellent businessman who had acquired many personal and business contacts and a modest amount of capital since moving to Texas.[26]

In 1888, Kurth purchased a sawmill from Charles L. Keltys and Keltys's son-in-law, J. A. Ewing, on the Cotton Belt Railroad, about two miles west of Lufkin. A small company town called Keltys was connected with the mill, which Kurth leased and to which he moved his growing family. It was to be the home of the Kurth family for three generations. According to the original bill of sale, the price paid for the Keltys properties was $11,000 in cash and the rest in short-term notes.[27] A considerably larger mill than the one he had operated in Polk County, it had a capacity of about 40,000 board feet a day. It consisted of a steam engine and boiler and the sawmill proper, with a circular saw, a gang saw, and an edger; a shed and yard with a number of lumber buggies; and about twenty yoke of oxen, along with yokes, chains, and long wagons. Also included were one planer, one Shay tram engine, and four tramcars.[28] A separate contract transferred about five thousand acres of timberland to support the mill operation. Kurth estimated that this acreage, together with the additional stumpage and land that he could buy, would keep the sawmill in production for seven or eight years.[29]

The mill required extensive repairs and improvements, however, and Kurth needed to build a list of steady customers. The next few months found him busily arranging orders for such major customers as William Cameron, the Palo Pinto Coal Mining Company, and various Kansas City retail yards. At the same time he was supervising the repair of the planing mill, seeking a secondhand boiler and steam engine for it, considering building a dry kiln, and shopping for a used trimmer. In addition he was trying to establish credit with Houston and Galveston bankers and haggling with the Cotton Belt officials over more satisfactory freight rates.[30]

[26] Conversation with Simon W. Henderson, Jr., August 10, 1964. The senior Henderson married Louise Reed, of Lufkin, in 1903 (his first wife had died in 1888 after a marriage of only seven months). His son, Simon W. Henderson, Jr., in time joined him in his business enterprises.

[27] J. H. Kurth to William Cameron, March 23, 1888; bill of sale, March 10, 1888, ACLCo. Papers, Forest History Collections, SFA Library.

[28] Bill of sale, March 10, 1888, ACLCo. Papers; *Lufkin Daily News*, October 30, 1955.

[29] Kurth to St. Louis Southwestern Railway, February 16, 1889, ACLCo. Papers.

[30] Kurth to Wm. Cameron & Company, May 22, 1888; Kurth to Palo Pinto Coal Mining Company, May 31, 1888; Kurth to E. T. Cowan Lumber Company, December 2, 1888; Kurth to George M. Dilley and Son, May 21, 1888; Kurth to

Within the year Kurth found that this extended operation demanded more capital and managerial supervision than he alone could provide. Thus in October, 1888, Simon W. Henderson joined him to form a partnership under the name Henderson and Kurth. Henderson initially invested about $8,000 in the company; later he closed his business in Corrigan and put his entire fortune in the sawmill. The partnership freed Kurth to spend more time on the road soliciting business and opened to the partners the credit resources and contacts that Henderson had developed in his mercantile business in Corrigan.[31] The partnership lasted only a little more than a year. In the fall of 1890, Henderson and Kurth joined with Sam Wiener, originally from Mississippi, to form the Angelina County Lumber Company (ACL Co.). In the new organization Kurth was president and general manager, Henderson was vice-president, and Wiener was secretary-treasurer and in charge of the office. At the same time Wiener's younger brother, Eli, became a clerk in the company's office.[32]

The foursome (Eli later became an active partner also) was as dissimilar a group as can be imagined. Kurth was German, well educated, somewhat aristocratic, abrasive, and at times blunt to the point of rudeness. Henderson was a Georgia cracker, with little formal education, but personable and practical. Sam and Eli Wiener were Jewish, good merchants, careful keepers of accounts, and hard workers. These four joined to form a corporation that eventually became a multimillion-dollar enterprise and lasted well beyond the lives of its founders.[33]

The status of the new company was indicated by a memorandum to the firm P. J. Willis and Brother, of Galveston, which acted as banker and wholesaler for the ACL Company. The memorandum listed the value of the sawmill and planer at $15,000; the tramroad, including engine and logging outfit, at $10,000; merchandise at $4,000; and lumber on hand at $3,500. Accounts due totaled $3,500. Land and timber holdings, including about 5,000 acres recently purchased, were listed at a total of $11,737. Total liabilities, including notes on recently acquired land, were $11,687. This left a net worth of $37,550, which did not indicate a large corporation even for East Texas in 1890.[34]

The next two decades were spent in a struggle to establish the ACL Company on a firm financial basis with a good credit rating in the South and the Middle West. An effort to operate a retail lumberyard at the ephemeral town of New Birmingham was discontinued. Orders were sought and filled as far afield as Saint Louis, Kansas City, and Chicago. Although sales were good and prices favorable in the early 1890s, the lumber business was marked by rather violent seasonal shifts. The company also waged a continuing battle with the officials of the Cotton Belt Railroad over freight cars and rates. The Cotton Belt was a narrow-gauge road until 1899, which meant that cars loaded at Keltys had to be reloaded at Tyler for transshipment.[35] The Panic of 1893 and the en-

Tyler Lumber Company, June 1, 1888; Kurth to L. T. Day, St. Louis, Arkansas and Texas Railway, May 31, 1888; Kurth to Halff and Neborn, March 24, 1888, ACLCo. Papers.

[31] Simon Henderson to Kurth, November 8, 1888; Henderson and Kurth to Wm. Cameron & Company, October 29, 1888; Kurth to Henderson, November 23, 1888, ACLCo. Papers.

[32] Sam Wiener to Mssrs. Focke, Wilkins, and Lange, October 5, 1890; Sam Wiener to Henderson, October 3, 1890, ACLCo. Papers.

[33] Lufkin Daily News, June 16–17, 1930; Beaumont Enterprise, August 21–22, 1957; Ernest Balch, "Biographical Sketches of Industrial Leaders in Lufkin and Angelina County" (master's research paper, Stephen F. Austin State College, Nacogdoches, 1949); conversation with Ernest L. Kurth, Sr., August, 1958.

[34] Memorandum of Agreement by P. J. Willis and Bro., Galveston, and Henderson and Kurth, Keltys, March 6, 1890, ACLCo. Papers.

[35] Kurth to Sam Wiener, March 28, 1889, November 11, 1889; Kurth to Fred Schuarte, St. Louis, October 24, 1890; Kurth to Louis Werner Sawmill Co., St. Louis, May 24, 1900;

suing depression hit the young company very hard. Prices dropped, and orders were exceedingly scarce even at lower prices. Kurth was forced to close the sawmill until the company could dispose of unsold stock, estimated at about $45,000. Like many other businessmen, they rejoiced at the election of William McKinley in 1896 and predicted that trade would improve by the end of the year. By the end of 1897 business was booming again, and the ACL Company had more orders than it could fill.[36]

The ACL Company also had to cope with what appeared to be more than the normal hazards of a hazardous business simply to keep the plant in operation. Within a year after they began operations, they lost the original dry kiln by fire, and in 1893 the boiler and power plant for the planing mill were destroyed. In 1898 the company was forced to build a new building and move the machinery into it. Kurth commented that the "old one was rotten and falling down." In 1901 the dry-kiln boiler exploded, and drying operations had to be halted until a new boiler could be procured and installed. In this disaster one of their employees was killed. In 1906, two years after a shutdown for extensive repairs and improvements late in 1904, the main mill burned (the stocks of lumber and most other equipment apparently were saved).[37] Nevertheless, by early 1907 the ACL Company was back in operation, and its owners were optimistic about the future of their business.[38]

In 1900, Kurth chartered his tramroad as a common carrier under the name Angelina and Neches River Railroad. This little short line ran east and north across the Angelina River and eventually was extended to Chireno, a small village near the Attoyac River in Nacogdoches County. In addition to using the main line as a stem for logging operations, the company ran a "mixed train daily" serving Dunnagan Station, Etoile, and Chireno. The completion of the A&NR also gave the ACL Company a more competitive position regarding freight rates. The company could now reach the Houston, East and West Texas on its own tracks and entirely bypass the Cotton Belt if desirable. Thus the A&NR could demand a division of the tariffs for its part of the long haul. The railroad was a valuable addition to the ACL Company properties and proved profitable both to the parent company and to itself.[39]

The ACL Company continued to prosper and expand in the twentieth century. In 1900 the company's financial statement listed its net worth at $160,000. By 1912 its net assets had grown to $1,262,000. Its timber holdings in Angelina, Nacogdoches, and San Augustine counties were by then numbered in the hundreds of thousands of acres.[40] In addition,

Henderson to A. S. Dodge (Cotton Belt), August 4, 1894; Henderson to R. H. Bowron (Cotton Belt), December 7, 1900, ACLCo. Papers.

[36] Kurth to Arkansas Milling Co., October 23, 1893; Kurth to Messrs. Dalsheimer and Worms, October 31, 1893; Kurth to Davis, Rosenberg, and Levy, October 31, 1893; Kurth to Chicago Belting Co., November 12, 1896; Kurth to Temet-Stripling Shoe Company, September 6, 1897; Kurth to August Gast Bank Note and Lithograph Co., September 28, 1897, ACLCo. Papers.

[37] Henderson and Kurth to A. C. Petri & Brother, January 2, 1889; S. Wiener to S. Hermsheim Bros. & Co., October 31, 1893; ACLCo. to A. E. Dodge, August 8, 1898; ACLCo. to A. Baldwin & Co., February 2, 1901; ACLCo. to Frost Lumber Co., January 30, 1905; ACLCo. to Wallace Pratt, August 3, 1906; ACLCo. to Texas and Louisiana Lumber Co., August 7, 1906, ACLCo. Papers.

[38] When completed, the new mill was a double-carriage rig with both a band saw and a circular saw, capable of producing 120,000 board feet of lumber a day. New safety features were included in the structure. ACLCo. Papers.

[39] Maxwell, *Whistle in the Piney Woods*, pp. 47–50, 60–61; ACLCo. to J. W. Allen, December 3, 1900; ACLCo. to L. J. Storey (railroad commissioner), December 17, 1900; ACLCo. to E. L. Sargent, December 21, 1900, ACLCo. Papers.

[40] ACLCo. financial statements, 1899–1912, (box 879), ACLCo. Papers. In 1916 for reasons not readily apparent the owners transferred the assets of the Angelina County Lumber Company to the San Augustine Lumber Company but continued to operate the original company as a sales outlet. Then in 1923 the name was changed back to Angelina County Lumber Company. Through all the changes the same group—the Kurths, the Hendersons, and the Wieners—continued to operate the enterprise. Conversation with S. W. Henderson, Jr., August 10, 1964; conversation of Otis W. Jayroe with John W. Cordray, June 27, 1963.

Ernest L. Kurth. *Authors' collections.*

Kurth, Henderson, and both Wieners held substantial interests in several other enterprises, including the Lufkin National Bank, the Lufkin Light Company, and the Lufkin Foundry and Machine Company.[41]

By 1900, Kurth had become a prominent figure in East Texas. Of medium height and stocky build, he wore a moustache and goatee and carried himself in a "princely and courtly" manner. He enjoyed well-fitting, carefully made clothing accented by unusual ties and stickpins. He had become an American citizen and was an active member of the Republican party. He strongly supported McKinley and campaigned for Roosevelt both in 1904 and 1912. By that time he had become a leader

in state politics and a person to be consulted about federal appointments during the McKinley-Roosevelt-Taft era. At various times Kurth served as a member of the Texas State Republican Executive Committee, chairman of the Republican party in Angelina County, a delegate to the conventions in Angelina County, a delegate to the Republican national convention, and a presidential elector. In 1924 he was his party's candidate for the office of lieutenant governor of Texas, and he made a statewide campaign as the running mate of Dr. George C. Butte against Miriam A. Ferguson and Barry Miller. He ran a creditable race but was decisively defeated. He did, however, carry his home county, which was normally strongly Democratic.[42]

Following his graduation from Southwestern University (Georgetown, Texas) in 1905, Ernest L. Kurth, the second son of the founder of the ACL Company, became associated with the Angelina County Lumber Company and other family businesses. The younger Kurth served as bookkeeper, sales manager, and then general manager of the lumber company. Upon his father's death in 1930, he became vice-president and then president in 1936, when Eli Wiener retired from active management. Under Ernest Kurth the ACL Company expanded into a great complex of enterprises. He was a

[41] Conversations with E. L. Kurth, Sr., 1955–59. William C. Trout, a practical mechanic and inventor, had worked in machine shops and sawmills in the North before moving to Lufkin in 1902. He was cofounder and general manager of the Lufkin Foundry. The company manufactured the Trout set works (for sawmill carriages), which he invented, and the Martin grip hook (for loading logs), which was also invented in the foundry. Later the foundry became important in the manufacture of oil-field equipment. Conversation with A. E. Cudlipp, March 8, 1960; conversation with S. W. Henderson, Jr., August 10, 1964.

[42] Kurth to C. A. Lyon, February 18, 1897; Petition to the Honorable William McKinley [copy], February 17, 1897; R. B. Hawley to Kurth, May 14, 1902; Lyon to Kurth, September 9, 1907; Kurth to G. W. Easum, March 20, 1912; Kurth to E. G. Christian, March 19, 1912; R. B. Creager to Kurth, February 18, 1925; Kurth to Creager, March 7, 1925, ACLCo. Papers; *Lufkin Daily News*, October 31, November 10, 1924. See also *Texas Almanac, 1926.* Joseph Kurth was also active in local civic and religious affairs. Although he had been reared a Roman Catholic, after his marriage he was a vigorous member of the Methodist church and largely built the Methodist church in Keltys. He contributed generously to support Southwestern University, a Methodist school, and sent his sons there. He gave land for the new school in Lufkin and for many years was a trustee of the local school district. He was a thirty-second-degree Mason, a Shriner, an Elk, and a member of the lumber manufacturers' fun organization, the Hoo Hoo. In his will he left funds for a library in Lufkin, now known as Kurth Memorial Library. Indeed, there was scarcely a civic enterprise in Lufkin or Angelina County in which Joseph Kurth did not participate.

cofounder of the Southland Paper Company, which in 1940 began producing newsprint from southern pine pulpwood, having built the first such mill in the South. Like his father he was active in state and national politics, generous in philanthropies and good works, and prominent in lumber and forestry associations. So great was his prestige in the economic life of the Piney Woods that one governor hailed him as "Mr. East Texas."[43]

This family business, which Joseph Kurth,

Sr., began on a modest scale in 1888, grew into one of the giants of East Texas lumbering. Under the direction of the Kurth family; Simon Henderson and his son, Simon Henderson, Jr.; and Sam Wiener and his brother, Eli, the ACL Company enjoyed continuing excellent management. No less than three generations of Kurths had directed the company and its related enterprises when the company was finally liquidated in 1966. Its growth and development are one of the sagas of the Texas story.

The Knox Family

Great wealth, a beautiful woman, and mysterious death are always prime ingredients of an exciting tale. Certainly for romantic appeal or human interest no fictional mystery can match the story of the Knox family in the Piney Woods of East Texas. William Hiram Knox, Sr., the founder of the line, was an experienced lumberman from the white-pine forests of the Great Lakes states who moved to Texas sometime before 1900.[44] He first lived in Texarkana, where he operated a lumber business, and then established a large retail lumberyard (described as the first such venture) at Mission, Texas, in the Rio Grande Valley. Sometime before 1903 he purchased a large timber acreage and built a sawmill plant east of Livingston, in Polk County. He also built and had chartered a common-carrier railroad, called the Livingston and Southeastern, which connected his mill with the Houston, East and West Texas at Livingston. Here the Knox family lived and operated a large lumber manufacturing business for at least a dozen years before 1913.[45]

It was apparently a violent family. The se-

nior Knox brought with him to Polk County from Texarkana and the North a reputation for direct action. Veteran lumbermen from the Livingston area recall that one of the Knox sons was killed in a free-for-all at an all-night dance hall. In another incident Knox senior reputedly shot down a white man on the streets of the mill town because the man had abused one of his black employees. Because of his position and the state of law and order, no one questioned Knox about the alleged murder.[46]

During an illness that struck while he was living in Texarkana, Knox was nursed by a friendly young woman who made herself so indispensable that she became a member of the family. A few years later she and W. Hiram Knox, Jr., were married. Lillian Miriam Knox moved easily and naturally into the family business, serving at some times as secretary and at others as confidante. Apparently she soon dominated completely the rather unenergetic Hiram, Jr., and thoroughly pleased and captivated the elder Knox.[47]

As the timber on which the Polk County mill depended neared exhaustion, Knox purchased a tract of land (estimated at 25,000 acres) in Sabine County and began building a new, larger mill near Hemphill. In 1912, during its

[43] *Lufkin Daily News*, October 30, 1955; October 27, 1960; March 15, 1966; conversations with E. L. Kurth, Sr., 1955–59; conversations with Otis Jayroe, 1962–63.

[44] One national lumbering periodical spoke of Knox as a "pioneer Michigan lumberman." *American Lumberman*, June 8, 1918. Men who knew him, however, insist that he was from Wausau, Wisconsin. Conversations with Judge J. W. Minton, Hemphill, May 13, 1956; August 5, 1958.

[45] *American Lumberman*, June 8, 1918; Mrs. James Watson, *The Lower Rio Grande Valley of Texas and Its Builders*.

[46] Conversation with Judge J. W. Minton, Hemphill, May 13, 1956; conversation with N. B. Weatherford, Camden, October 13, 1959.

[47] Conversations with Judge J. W. Minton, Hemphill, May 13, 1956; August 5, 1958.

Lillian Knox in formal pose. *Authors' collections.*

construction, the senior Knox died, but Hiram completed the project, and the mill began operation in 1913. By all descriptions the new Knox mill was a modern double-band rig, with dry kilns, planers, and other auxiliary equipment necessary to produce high-quality lumber, and employed 400 to 500 workers. East Mayfield, the Knox mill town adjoining Hemphill, was a "model company village" with better-than-average houses, churches, a school, a theater, and a small hospital. Knox bought more land until he owned approximately 50,000 acres, and built a short-line railroad to connect with the Santa Fe at Bronson. The road, chartered in 1912 as the Lufkin, Hemphill, and Gulf Railway, provided a means of ingress to East Mayfield and an outlet for the lumber products of the Knox mill. Local wags asserted that LH&G stood for "Left Hemphill and Gone" and also for "Lillian, Hiram, and Gussie" (the family cook). For his wife and six children Knox built a large, commodious home, called a "mansion" by the local inhabitants.[48]

In this remote and rural Eden, Lillian Knox played the role of Lady Bountiful. She furnished her home with costly furniture, tapestries, and imported china. One room was given over to an "art gallery," and a large cellar provided various vintage wines for guests. In a park near the house she maintained deer, turkeys, ducks, and a drove of peacocks. On occasion she imported theatrical companies and musical groups to perform in East Mayfield. For these events she habitually distributed tickets free to company employees and friends in Hemphill.

The entire region came to know her generosity and largesse. At Christmas time her gifts ran to hundreds if not thousands of dollars, and she reputedly made it her goal that no family was without a "bird" for Christmas. Throughout the year she was an angel of mercy

to the poor of the county. She was known to sit up all night nursing a sick child or woman and to press cash or checks for considerable amounts into the hands of destitute parents. On more than one occasion she was said literally to have given the coat off her back to a needy woman or girl. Labor troubles were at a minimum at East Mayfield.[49]

The coming of World War I in 1917 brought a boom in shipbuilding on the Gulf Coast and a great demand for ship timber (see chapter 12). The forests of the Knox Lumber Company were well suited to the manufacture of large, heavy timbers, and business soon flourished as the company acquired a number of government contracts. Although the shipping-board orders were attractive, they also carried deadline stipulations and penalties for failure to deliver promised timbers. The Knox Lumber Company, under the direction of Hiram Knox, evidently was ill-prepared to handle such contract work, and soon there developed a critical shortage of logs, threatening loss of the government contracts and even financial ruin from the penalties imposed for failure to meet agreed quotas. According to the best accounts, Hiram felt that the situation was hopeless and was ready to concede defeat.

At this point Mrs. Knox appeared in a new role. She wagered her husband that if he turned over the logging operations to her and confined his efforts to improving the efficiency of mill operations she would provide all the logs needed to fulfill the government contracts. He agreed, and she donned breeches and jacket, mounted a horse, and for weeks regularly went into the woods to supervise the logging and transportation of the logs to the mill. She filled up the millpond, the rail sidings, the docks, and all other available spaces. In fact, she covered up the mill with logs, until her husband begged her to pull back until he could catch up.

[48] According to the press, Hiram Knox inherited about $10 million upon his father's death. No doubt a large part of the estate was the East Mayfield property. *Houston Post*, January 2, 1923.

[49] Conversations with Paul Wilson, Sabine County, August 5–6, 1958; conversation with W. T. Pulliam, Hemphill, August 18, 1958; conversation with Judge J. W. Minton, August 5, 1958; conversation with Floyd Smith, Hemphill, May 1–10, 1958.

She won the bet, and the company met its contracts for ships' timbers in good time.[50]

Mrs. Knox was also promoting the war effort in other ways. Under her leadership company employees raised more than 150 acres of
war gardens and built a canning factory to preserve the produce. Every employee of the Knox
Lumber Company was a subscriber to Liberty
bonds and war savings stamps, and the Red
Cross numbered twelve hundred members
from Knox Company families. During the influenza epidemic of 1918, Mrs. Knox organized nursing teams and aides for the stricken
population. Truly she was Lady Bountiful, or
even Lady Miraculous, to the people of Hemphill and Sabine County.[51]

The drop in land and lumber prices after the
war placed Knox in serious financial difficulties, for he had purchased additional lands on
credit at inflated prices during the war years. In
1921–22, after negotiating with several prospective buyers, including Ernest Kurth, of the
ACL Company, Knox sold his mill and most of
the timberlands and part interest in the railroad to T. L. L. Temple and the Southern Pine
Lumber Company.

Although still wealthy and holding lands
in East Texas, the Rio Grande Valley, and Mexico and some oil interests, Hiram Knox was
depressed and tended to spend much time
alone.[52]

In the early-morning hours of November 26,
1922, he was found shot to death in his home
in Hemphill. He held an automatic pistol
loosely in his hand and had letters addressed to
his lawyer and his mother in his coat pocket.

Lillian Knox in later years. *Courtesy of Mrs. Frank
Hathcock, Nacogdoches.*

Apparently death was caused by a single bullet
wound in the back of his head. Mrs. Knox was
said to have borne up bravely at the funeral
and to have looked charming, though pale,
dressed in severe black.[53]

To many it was a clear case of suicide. To
others it only appeared to be suicide but actually was a cleverly executed murder. The sheriff of Sabine County asserted that it was odd
that Hiram Knox's hair was not singed or
burned as it would be if the pistol had been
held at his head and fired. He insisted that the
pistol would have to be held at least a foot
away from his head to leave no powder burns
and that this was virtually impossible for Knox
alone to accomplish. Furthermore, he said, the
bathroom window was open, and tracks of

[50] Conversation with Judge J. W. Minton, August 5, 1958;
conversations with Paul Wilson, August 5–6, 1958.

[51] *American Lumberman*, June 8, 1918; Watson, *The Lower
Rio Grande Valley of Texas and Its Builders.*

[52] Ernest Kurth insisted that on at least two occasions he and
Hiram Knox had agreed on terms for the purchase of the Knox
properties, only to have Mrs. Knox veto the agreement the next
morning. Kurth thoroughly distrusted Mrs. Knox and believed
the worst of her. "She looked like an angel," he said, "but she
was a devil." Conversation with Ernest Kurth, Sr., August,
1959; *Houston Post*, November 30, 1922; R. E. Minton, "Trib-
ute to the Temple Family," Temple Industries Papers, Forest History Collections, SFA Library.

[53] *Dallas Morning News*, November 27, 1922; *Redland Herald* (Nacogdoches), November 30, 1922. The letters were later
described as "inconclusive."

men's shoes were clearly visible in the soft ground outside. There were other minor discrepancies, such as an unexplained spot of blood and the position of the hands and arms of the deceased, which made the suicide theory open to question. The family chauffeur announced that a few days before Hiram's death he had driven Mrs. Knox to another county, where she met and talked with a strange man. Other persons came forward saying that they too had seen her talking with strange men. Finally a rumor, supposedly originating in the local post office, spread that Mrs. Knox had received in the mail the black dress that she wore to the funeral the very day that Hiram Knox was shot. Obviously, according to gossip, she had ordered it several days earlier.[54]

The coroner returned an "open verdict," and the question stewed for several weeks. Finally, on January 2, 1923, the Sabine County sheriff arrested Mrs. Knox, charging her with murder or conspiracy to murder. At the examining trial the testimonies of the now former sheriff, the chauffeur, and others were read into the record, and Mrs. Knox was bound over to the grand jury. A haggle developed over the amount of bond to be set, and the attorneys finally agreed on a bond of $5,000.[55]

After an intensive investigation the Sabine County Grand Jury declined to indict Mrs. Lillian Knox for the murder of her husband. In the parlance of the court, she was "no-billed."[56] That ended the legal prosecution but not the controversy. The people of Hemphill and Sabine County divided almost equally on the question of Mrs. Knox's guilt, and forty years later they still discussed the issue with considerable interest and heat. A large number of lumbermen, perhaps most, thought that she was guilty but had cleverly covered her tracks. Mrs. Knox moved away from East Texas, and her later life was a tragic one. According to friendly sources, she eventually became destitute. After two major fires the Temple Company discontinued operation of the great mill and dismantled it. The Lufkin, Hemphill and Gulf Railway was abandoned. Today all that is left of East Mayfield and the once-flourishing Knox lumber empire are some stone foundations and a few decaying and dilapidated shacks. Only the memory of the "Lady Bountiful of the Piney Woods" remains, and even that memory is shrouded in controversy.[57]

---

[54] *Daily Sentinel* (Nacogdoches), January 4, 1923; *Dallas Morning News*, January 5, 1923; *Houston Post*, January 3–4, 1923.

[55] According to various accounts, Mrs. Knox had been to Beaumont, where she had spent the Christmas holidays with friends, and was arrested when she returned to Hemphill. She refused the initial demand of $25,000 bail, saying that she was innocent and that bond in that amount was unreasonable. Apparently, until the judge set a permanent bond, a deputy sheriff

guarded her at her home. Despite conflicting newspaper stories, it is doubtful that she spent any nights in the county jail. *Dallas Morning News*, January 5–8, 1923; *Daily Sentinel*, January 5–8, 1923; *Houston Post*, January 5, 1923.

[56] *Daily Sentinel*, April 19, 1923; *Dallas Morning News*, April 19, 1923; *Houston Post*, April 19–20, 1923.

[57] Conversations with Paul Wilson and townspeople of Hemphill, August 5–6, 1958; conversations with Judge J. W. Minton, May 13, 1956, August 5, 1958; conversations with E. L. Kurth, Needham B. Weatherford, Clyde J. Woodward, Sr., Simon W. Henderson, Jr., and Clyde Thompson, Diboll.

CHAPTER 8

The Lumber Worker

The shanty is our station
and lumbering our occupation.—folk song

As mentioned earlier, the stranger who visited the pineries of East Texas or western Louisiana expecting to find an exotic, distinctive lumber worker as depicted in song and legend would have been disappointed. He would not have found the logger a hard-working, hard-drinking, free-living lumberjack, dressed in a colorful shirt, strong boots, and tasseled toboggan. The sawmill hand did not give the appearance of a neat, skilled mechanic. The southern lumber worker was by no means Bunyanesque. In dress, language, racial background, and family habits he was very similar to the East Texas farmer or small-town laborer. Yet he was a distinct type. The usual lumber worker did not shuffle back and forth between sawmill work and farming.[1] Most often he made a career of lumbering—until no more work was available. As one veteran sawmill man remarked, "When pine resin gets into your blood it's hard to get out."[2]

The typical lumber worker was not only a native-born citizen of the United States but southern-born and most often Texas-born. Of 66 company employees who had risen above the level of common laborer in one Texas corporation, all were white American citizens. Of these, 40 were Texas-born, and 16 had been born in other southern states. The remainder included 7 from northern or eastern states, 2 from Canada, and 1 from Europe.[3] According to the United States census of 1910, of more than 22,000 male workers in the Texas industry in all categories, approximately 10,000 were listed as native-born white, 9,000 as native-born black, and about 1,650, or 7 percent, as foreign-born.[4] These figures are representative of the general breakdown of lumbering employees during the entire half century under study. The proportion of blacks was somewhat smaller in 1880 and during the war years 1917 to 1919. Throughout the whole period, however, the ethnic origins of East Texas lumber workers remained surprisingly stable. Of 100 lumber workers in any of the larger companies, 50 were likely to be native-born whites, more than 30 of whom were Texas-born; 40 were American-born blacks, most of whom were native-born Texans; and the remaining 10 included 5 to 7 foreign-born workers, a few American Indians, and others. Thus 9 out of 10 were native-born Americans, two-thirds were native Texans, and most of the others were southerners. Four out of 10 were black. It was a remarkably homogeneous labor force.[5]

[1] Conversation with Needham B. Weatherford, Camden, October 13, 1959, Oral History Collections, Forest History Collections, Stephen F. Austin State University Library, Nacogdoches; hereafter cited as SFA Library. Unless otherwise noted, interviews and conversations cited are in these collections, and all interviews were conducted by principal author Maxwell.

[2] Conversation with Clyde J. Woodward, Sr., Nacogdoches, October 3, 1967.

[3] "The House of Thompson," American Lumberman, September 26, 1908.

[4] U.S. Department of Commerce, Bureau of the Census, United States Census, 1910, Supplement for Texas, table 7.

[5] Ruth A. Allen, East Texas Lumber Workers: An Economic and Social Picture, 1870–1950, pp. 51–59.

A group of Thompson employees, ca. 1908. *Courtesy of Stephen F. Austin State University, Forest* *History Collections, Thompson photo albums.*

The social, economic, cultural, and religious backgrounds of the workers were also homogeneous. It is not surprising that most lumber workers came from farm families, especially since as late as 1900 about two-thirds of all American families lived on farms. Most of these farms were small and poor, and the migration to the sawmill and logging camp represented an opportunity to earn more cash money than was possible in agriculture. The educational level was low but comparable to that of other poor rural southern communities at the turn of the century.[6] Typically, the worker was a nonreader, contenting himself with the weekly newspaper and, rarely, a cheap book. Most workers enjoyed a common religious heritage: Protestant, evangelical, and fundamentalist with a strong emphasis on a direct personal religious experience. The Baptist and Methodist denominations predominated, with a smaller proportion of Presbyterians and other Protestants. Roman Catholics were few and regarded with suspicion, and Episcopa-

[6] See chapter 9 for a more detailed discussion of education in the company town.

lians were practically unknown. With this common religious heritage the Texas sawmill workers held similar theological, ethical, and moral standards. At the same time they accepted and condoned the same pattern of lapses from this rather rigid and puritanical code. Individual behavior varied greatly, and some company executives were more severe than others in enforcing their own moral codes. The East Texas lumber worker typically drank, albeit not gracefully or before his family. He gambled but nodded in approval to sermons denouncing gaming. He gave lip service to the sanctity of marriage but often was adulterous when the opportunity offered. Having little private property himself, he frequently had scant regard for property vested in others, especially absentee landlords. Despite the efforts of the clergy and millowners, the worker himself, as distinct from his family, was not a habitual or regular churchgoer. When or if he thought of things theological, he was narrow, literal, and often bigoted. In these attitudes he did not differ markedly from his nonsawmill neighbors in the southwestern United States.[7]

The typical sawmill worker was a career lumberman, but there were many exceptions. At any given time in a large mill there were some casual workers who tried sawmilling for a while and then drifted on to something else. Increasing numbers of young men, often from second-generation sawmill families, grew up around the mill and began their work there but, seeing little chance for advancement, got out while they were still young and single. Many business and professional men in Houston, Dallas, and Shreveport can claim brief sawmill or logging-camp experience in their youth. Of course, many a young man began working in a sawmill as a temporary occupation but found himself there years later. Managers and owners often commented that it was unlikely that a worker who had put in as much as ten years in the lumbering industry would turn to something else.[8]

Many farmers were seasonal sawmill or logging workers. They were a source of labor that could be depended upon in the late fall and winter, especially if business was booming and the mill needed extra hands. Many farmers preferred to work in the woods, bringing their own teams, for which they received extra wages. Other farmers owned small portable mills and periodically engaged in sawmilling themselves, particularly when lumber was bringing a good price. These so-called peckerwood mills appeared and disappeared as the business cycle in the industry changed. Often the farmer and his sons were owners, managers, loggers, sawyers, and laborers all rolled into the same three or four persons.

Perhaps the most usual kind of farmer-lumberman was the contract logger. He cut his own timber and delivered the logs to an agreed point—the railroad siding, the millpond, or, in the early days, before 1900, perhaps the bank

of the Sabine or the Neches River. Some lumber companies depended on contract loggers for their log supply. Other farmer-loggers cut and hauled company timber under the direction of the woods superintendent. They often called themselves contract-loggers also.[9]

The most common job at a sawmill was that of laborer. This was the standard unskilled manual-labor job that drew the basic rate of pay. In 1890 the prevailing rate of pay for common labor at the Angelina County Lumber Company was $1.25 to $1.50 a day for an eleven-hour day. In 1897 the Alexander Gilmer Company paid its unskilled employees $1.25, $1.50, and $1.75 a day for an eleven-hour day. In 1905 the payroll of the Foster Lumber Company at Fostoria disclosed that the great majority of its employees were paid $1.40 to $1.50 a day for ten-hour days. In 1914 the Speek investigation (for the U.S. Commission on Industrial Relations) of working conditions in three Texas lumber towns reported that the Gilmer Lumber Company was paying $1.55 to $1.75 for unskilled labor for a ten-hour day; the Kirby Lumber Company paid $1.50 to $1.95 for a ten-hour day to its unskilled workers, who comprised 260 of a 350-man work force.[10]

From this sampling it is apparent that in the twenty-five years from 1890 to 1914 wages for unskilled labor did not advance appreciably, though the workday was reduced from eleven to ten hours. A great variety of jobs and duties came under the term "common labor," many demanding some knowledge of sawmill machinery and most requiring considerable strength and coordination. The men behind

---

[7] Interviews in Oral History Collections, Forest History Collections, SFA Library.

[8] Conversation with Needham B. Weatherford, Camden, October 14, 1959; conversation with Clyde J. Woodward, Sr., Nacogdoches, October 3, 1957.

[9] Conversation with H. Z. Collier, Kennard, July 29, 1964; conversation with M. S. ("Pud") Morris, Sabine County, August 5, 1958; conversation with George Morrison, Huntington, August 6, 1958.

[10] Mill and Yard Book, 1890, ACLCo. Papers, Forest History Collections, SFA Library; Time Book, January–September, 1897, Alexander Gilmer Papers, University of Texas Library, Austin; Payroll for 1906, Foster Lumber Company Papers, Forest History Collections, SFA Library; Peter A. Speek, "Notes on Investigations of Three Texas Lumber Towns," U.S. Department of Labor, Reports of the Commission on Industrial Relations, 1914 (microfilm), Forest History Collections, SFA Library.

the saws, the slab men, the hog feeders, and the men who worked in the bear pit were all classed as laborers. In the yard lumber stackers, sorters, dolly pushers, dry-kiln feeders, and waste burners were also grouped as unskilled laborers. All but a half-dozen men in the planing mill were so listed as were workers at various jobs in the woods and those with the steel gang.[11]

Many workers in the mill and especially those in the woods were paid on a piece basis. Workers who made and stacked pickets, raised logs from the mill pond or the river, or made ties; fallers ("flatheads"); and buckers were all regularly paid at a piecework rate. Foster Lumber Company records for 1905 indicate that 52 of slightly more than 400 employees were paid on a piecework basis. Payment by the piece obviously allowed strong, energetic workers to make more money a day than they would make working at a daily rate. Yet this self-imposed speed-up could not fail to take its toll in accidents and illness.[12]

The best job in the sawmill, apart from the manager's, was that of saw filer. He received wages of $5 to $10 a day and was constantly in demand. A Houston paper advertised in 1910 for a saw filer, offering $150 a month. He dominated the file room, which was usually on the second floor of the mill, where at least some of the mill noises were muted. With his helper the saw filer cared for the various saws in the mill—band saws, gang saws, circulars, and trimmers—kept them sharp and ready for use in their turn. His job was, however, frequently a terminal position. Generally he went no higher in the company. In an analysis of sixty-six middle-management personnel of one large lumber company only three persons listed the position of saw filer as experience leading to their current jobs.[13]

Other key skilled jobs in the sawmill ranged downward from sawyer to edgerman, gang-saw operator, and trimmer. In the yard the foreman and the men in charge of the dry kiln were the principal skilled workers. In the planing mill the foremen, engineers, assistant engineers, and graders were key personnel. In the sawmill itself the foremen were usually paid by the month, their salaries about 1910 ranging from $100.00 to $125.00. The sawyer averaged $5.00 to $6.00 a day; the others, from $2.00 to $5.00 a day, depending on demand, seniority, and skill. Edgermen averaged about $3.00; trimmers, file assistants, graders, deck sawmen, and hookers, about $2.00 to $2.50. Carriage men's salaries ranged from $1.75 to $2.50 a day. Outside the mill such skilled workers as loaders, mule skinners, bull punchers, railroad engineers, chief boiler firemen, machinists, plant engineers, and blacksmiths averaged $2.50 to $3.50 a day. As a generalization, no more than one-fourth of the men at any given mill were rated as skilled labor or drew more than $2.00 a day. Like the wages for common laborers, the wage scale for skilled workers does not appear to have changed markedly between 1890 and 1914.[14]

In terms of annual earnings the take-home pay was considerably less than the amount the worker might anticipate. In a year of 365 days, after deducting 52 Sundays, 3 or 4 holidays, and 7 to 10 days off during mill repairs, the worker might expect the mill to be in operation an average of 300 days a year. Very few mills operated 300 days in any year, however. The usual operation seems to have been between

[11] Conversation with Needham B. Weatherford, Camden, October 13, 1959; conversation with Clyde J. Woodward, Sr., Nacogdoches, October 3, 1957; Woods and Mill Pay Books, ACLCo. Papers.

[12] Payroll for 1905, Foster Lumber Company Papers; Allen, *East Texas Lumber Workers*, pp. 73–74.

[13] Speek, "Notes on Investigations of Three Texas Lumber Towns"; Allen, *East Texas Lumber Workers*, p. 75; conversation with Ed Bird, Camden, July 23, 1958; conversation with Simon W. Henderson, Keltys, August 10, 1964; *Houston Chronicle*, March 8, 1910, quoted in Harry Weaver, "Labor Practices in the East Texas Lumber Industry" (master's thesis, Stephen F. Austin State University, Nacogdoches, 1961); *American Lumberman*, September 26, 1908.

[14] Conversation with Clyde J. Woodward, Sr., Nacogdoches, October 3, 1957; conversation with Ed Bird, Camden, July 23, 1958; conversation with N. B. Weatherford, Camden, October 13, 1959; conversation with Simon W. Henderson, Keltys, August 10, 1964; Weaver, *Labor Practices*, pp. 31–33; Allen, *East Texas Lumber Workers*, pp. 99–100.

260 and 280 days, with some mills falling below that.[15] If the mill did not operate, the worker was not paid.

The causes and reasons for shutdowns were many and varied. Accidents and damage to plant machinery usually meant that the mill must suspend operations for several days at least. A long siege of wet weather would halt logging operations and force the mill to shut down until the log supply could be replenished. Fires were an ever-present danger to the worker's job. In the period between 1902 and 1906 a national lumber magazine reported no less than twenty-eight major fires involving Texas sawmills. These included fires in mills, dry kilns, lumberyards, and planing mills, the destruction of any of which would shut down entire plants. It was also standard practice for owners to shut down mills in slack times, when they were overloaded with hard-to-move inventories or when prices were depressed. In such instances each worker lost some of his projected annual income.[16]

Even when the mill was in operation, there was great variation in the number of days that a man might work in a given month. The records of both the Gilmer Lumber Company and the Angelina County Lumber Company indicate that just because a man was on the payroll was no reason to suppose that he would work full time or draw full wages. For example, in a typical month (May) a woods crew might work eleven to twenty-five days. Members of mill and yard crews might work ten to twenty-six days. In the fall of 1914 the Kirby Lumber Company, at Kirbyville, operated only five days a week, a common work week, according

to office personnel, in "slack times."[17]

For the average worker advancement in the lumbering world was slow and uncertain. Even in a large mill there were one saw filer, no more than three or four sawyers, two edgermen, one trimmer, one mill manager, and one assistant manager, who often doubled as extra sawyer. In the yard, the planing mill, and the woods and on the tramroad the ratio was about the same, and many skilled workers and mill managers continued in the same jobs for ten, fifteen, or even twenty years. Yet advancement did occur, and the records of the industry provide many instances of eager young men who worked up from common-laborer jobs to important managerial positions.[18]

An example was Clyde Woodward, who was born in 1888 in Moscow, Polk County, Texas. While he was still in his teens, he took his first job, with W. T. Carter & Brother Lumber Company, in Camden. He showed special talent for things mechanical and worked on the Carters' common carrier, the Moscow, Camden and San Augustine Railroad. In 1911 he moved to Nacogdoches and joined the Frost-Johnson Lumber Company. He worked for the company railroad, the Nacogdoches and Southeastern, as mechanic, locomotive engineer, and master car builder. When the logging camps were built or moved, he had charge of the operations. Eventually he rose to become manager of the N&SE and director of its operations. For a time he served as mill superintendent as well. To principal owner E. A. Frost and general manager H. Worth Whited, Clyde Woodward was a highly respected and trusted associate. He retired in 1952, when the Frost properties were sold to the Olin Company.[19]

Another example from about the same pe-

---

[15] A perusal of the ACLCo. Papers of the period 1889–1914 reveals how often the mill was shut down for one reason or another. See Sam Wiener to Louisville Jeans Clothing Co., December 18, 1891; Joseph Kurth to A. L. Clark, June 15, 1896; Kurth to Wm. Cameron, February 13, 1889; Kurth to Wm. Cameron, August, 1906, ACLCo. Papers; *American Lumberman*, August 11, 1906; Allen, *East Texas Lumber Workers*, pp. 99–100.

[16] See J. Kurth to A. L. Clark, June 15, 1896, ACLCo. Papers; *American Lumberman*, 1902–1906; *Southwest*, November, 1907.

[17] Time Books, January–September, 1897, Gilmer Lumber Company Papers, University of Texas Library, Austin; Time Books, 1888–1902, ACLCo. Papers; Speek, "Notes on Investigations of Three Texas Lumber Towns."

[18] Conversation with Simon W. Henderson, Keltys, August 10, 1964; see other biographical sketches in Oral History Collections, Forest History Collections, SFA Library.

[19] Conversation with Clyde J. Woodward, Sr., Nacogdoches, October 3, 1957.

Employees at a typical big mill lumber operation. The children frequently carried lunches to their fathers at the mill. *Courtesy of Stephen F. Austin* *State University, Forest History Collections, Thompson photo albums.*

riod illustrates the rise of a logger. George Morrison was born on a farm in Tennessee, where he grew to manhood. He worked in logging camps in Arkansas and then went to California, where he spent several years in the redwood camps. He returned to Arkansas and from there moved to Keltys, where he took a job with the Angelina County Lumber Company. He worked mostly in the woods as a helper, faller, mule skinner, bull puncher, and loader. Eventually he rose to be foreman and finally logging superintendent. When he retired, he had worked for Angelina about forty years. In excellent health, he lived until he was almost one hundred years old.[20]

Although company officials made much of employee loyalty and continued service, by no means a majority of lumber workers put in long careers with a single company. Peter Speek, in his investigation of Texas lumbering conditions for the United States Department of Labor, estimated that one-third of the work force was moving from one job to another at any given time. This rate varied greatly among companies, of course. The records of representative lumber companies show a considerable turnover among workers in a year's time. An analysis of one large lumber operation shows that its skilled and managerial personnel had worked on an average for two other companies. Only one-sixth of the employees had spent their entire careers with the company. The same study also indicated that opportunities for advancement were greatest for those employees who worked in the offices as clerks, stenographers, shipping clerks, and bookkeepers. Of the sixty-six key employees no fewer than twenty-nine listed their principal, and often their only, experience as office work. The same

[20] Conversation with George Morrison, Huntington, August 6, 1958.

Company office. *Courtesy of Stephen F. Austin State University, Forest History Collections,* *Thompson photo albums.*

Middle management officers with the Thompson Lumber Industries. *Courtesy of Stephen F. Austin* *State University, Forest History Collections, Thompson photo albums.*

group reported that thirteen had attended college at least one year and three had gone to business college. Many were currently directing operations in the mill, yard, planer, or commissary, on the tramroad, or in the woods. As in many other industries, a superior education, friendly personality, neat appearance, and proximity to and acquaintance with top management officials were often the stepping stones to advancement in the lumbering industry in the Gulf Southwest.[21]

For the black worker advancement was even more difficult and uncertain. In the segregated world of Texas and Louisiana lumbering in the generation before 1930, a black foreman or manager supervising both white and black workers was almost unknown. Indeed, there were frequent threats to bring in foreign workers to replace blacks in common-laborer jobs. Too often the problem was one not of advancement but simply of holding on to a job. As indicated earlier, the black worker was usually the lowest-paid, the last hired, and the first fired or laid off. Yet there were cases of black promotions. Scattered throughout East Texas were a few black woods bosses, steel-gang foremen, locomotive engineers, and sawyers. Perhaps the best job open to the black was that of woods sawyer, or flathead, for which he was paid on a piece-work basis. Most carriage men were also black, and, although their job was dangerous and exhausting, it was much sought after because it usually paid a higher wage than that of other jobs. In the great majority of cases, however, even though the job called for skill and superior strength (such as the job of the man in the bear pit) the black holding the job was listed simply as laborer.[22]

Accidents also took their toll of the worker's income and well-being, increasing his days and weeks of unemployment. They could shorten or even terminate his career in the lumber industry; they could end his life. Of all manufacturing occupations, logging and sawmilling were two of the most dangerous. The United States Department of Labor, Bureau of Labor Statistics, rated the lumber worker seven times as likely to "sustain an injury on the job" as an average of all other workers in manufacturing industries. Among United States industries logging and sawmilling ranked first and third in frequency of disabling accidents and first and fourth in number of fatalities. The record in Texas was no better than that of the nation as a whole.[23]

Accurate statistics on the number of accidents in Texas lumbering are difficult to assemble, especially for the earlier years. The companies were reluctant, for obvious reasons, to report annual totals. In company records occasional references can be found to accidents and injuries of specific workers, but complete lists are generally lacking. With the advent of workmen's compensation insurance, introduced in Texas in 1913, statistics became increasingly reliable. Toward the end of the bonanza era the National Council on Compensation Insurance issued a summary for the years 1922 to 1926. By that time statistics-gathering methods had become relatively efficient, and most companies had elected to participate in some form of workmen's compensation. For the years 1922 to 1926 the council reported that in the Texas logging industry, which had a total payroll of $26,362,300, there were 194 serious and 7,696 nonserious accidents. In sawmilling, with a payroll of $17,444,100, there were 108 serious and 4,231 nonserious accidents.

[21] Speek, "Notes on Investigations of Three Texas Lumber Towns"; Time Books, Angelina County Lumber Company, 1888–1912, Time Books, Foster Lumber Company, 1905–20; Time Books, Frost-Johnson Lumber Company, 1910–30, Forest History Collections, SFA Library; "House of Thompson," *American Lumberman*, September 26, 1908. It is interesting and perhaps significant that no less than twelve of these key employees had once worked for the Kirby Lumber Company but had left for other jobs.

[22] Allen, *East Texas Lumber Workers*, pp. 56–59, 86–87;

conversation with George Morrison, Huntington, August 6, 1958; conversation with Needham B. Weatherford, Camden, October 13, 1959; conversation with S. W. Henderson, Keltys, August 10, 1964.

[23] U.S. Department of Labor, Bureau of Labor Statistics *Bulletin*, no. 490 (1928); *Bulletin*, no. 667 (1939), quoted in Allen, *East Texas Lumber Workers*, pp. 109–14.

In the planing mills, with a payroll of $9,191,500, there were 37 serious and 1,335 nonserious accidents. In lumberyards, with a payroll of $29,613,100, there were 49 serious and 2,696 nonserious accidents. Totals for the four-year period were 388 serious and 15,958 nonserious accidents in the Texas lumbering industry.[24]

A specific example can be found in the workmen's compensation reports on the logging operations of the Angelina County Lumber Company for 1925 (no doubt these figures were included in the comprehensive figures given above). In that one year the company reported 17 logging accidents involving lost time. There were no fatalities, but time lost ranged from 1 day to 170 days, with an average of 22 days. The injuries were described as cuts, mashed limbs, fractures, sprains, and bruises. The total averaged about one and one-half lost-time accidents per month, and the lost time averaged more than three and one-half weeks. Clearly accidents were a constant and major hazard and took a heavy toll.[25]

In the early days the worker injured in an accident on the job had few means of recovering damages or obtaining benefits. Until 1913 his chief recourse was a suit against the company at common law, in which his chances of recovery were meager indeed. Not only did the company come to court well represented with legal counsel experienced in civil suits, but the three traditional defenses in common law against the employer's liability usually assured a verdict in favor of the company. If it could be proved that the worker had voluntarily assumed the risk of a hazardous occupation, if a fellow worker had through negligence contributed to the worker's injury, or if the worker himself had been negligent, the court habitually would find the company not liable. It was a common practice of Texas and Louisiana lumber companies to require the worker to sign an application form that included a statement that he was "fully physically and mentally competent to perform all duties connected with the said position, being fully experienced therein and throughly familiar with the duties thereof."[26]

By and large the injured worker preferred to depend on the benevolence of his employers rather than risk a suit at law. Such a suit would certainly antagonize the owner, and if it failed, the worker could expect nothing from the company. Many owners, including Gilmer, Kirby, Lutcher and Moore, and Temple, regularly saw to the care of injured employees, particularly those who had been employed for a number of years and were considered permanent. Company policies in regard to compensation for death or serious accident varied from owner to owner. Indeed, until 1913 any action was optional. According to a veteran employee of the Angelina County Lumber Company, when an employee was killed on the job, Joseph Kurth provided his widow with a company-owned house for the rest of her life and gave her a drawing account at the commissary for groceries and other household needs. On the other hand, at another mill in the same county, if a worker was killed in a mill accident, the manager not only denied the widow any compensation but charged her for the lumber for the coffin. These incidents probably represent extremes, but it seems clear that the action taken by the company was governed by the circumstances of each case—and the whim of the owner.[27]

[24] Serious accidents were classified as those resulting in death, permanent total disability, or major permanent partial disability. Nonserious accidents include all others that involved lost time beyond the waiting period. From memorandum of the National Council on Compensation Insurance, D. D. Smith, secretary, to author Maxwell, October 15, 1959, Forest History Collections, SFA Library.

[25] From Workmen's Compensation Reports, 1925, ACLCo. Papers. See "Mayhem in the Woods," *Gulf Coast Lumberman*, June 1, 1949, for a discussion of the pattern of more recent accidents in the logging industry.

[26] Elizabeth Brandeis, *History of Labor in the United States, 1896–1932*, 3:564–70; application form for employment, Angelina County Lumber Co., June 24, 1912, ACLCo. Papers.

[27] Weaver, *Labor Practices*, p. 45. See also Allen, *East Texas Lumber Workers*, p. 108, for other examples.

Although during the entire half century from 1880 to 1930 the lumbering industry in Texas employed more workers than any other industry, the employee had little political power to improve his working conditions. He was unorganized, socially isolated, and economically utterly dependent upon the company. In time came passage of employers' liability laws, the abrogation of the traditional common-law defenses, and eventually the enactment of workmen's compensation laws, all products of the spirit of reform that pervaded the country from the late 1890s to the outbreak of World War I. The Progressive Era, as this period of reform was called, produced social and economic legislation throughout the states as well as at the national level. Led by Wisconsin, New York, Oregon, and California, other state legislatures passed a significant number of laws benefiting industrial labor in such areas as wages, hours, on-the-job safety, anticoercion, and employers' liability. In Texas such popular leaders as James S. Hogg, Thomas Campbell, and James E. Ferguson claimed Progressive principles and spoke on behalf of the workingman. With the arousing of the social conscience of the state, other political figures could do no less.

The Texas State Federation of Labor (AFL) agitated and lobbied for legislation for the betterment of working conditions and laboring standards. Although the AFL had no organization in the Piney Woods, its goals would benefit the lumber workers as well as its own members. Thus, although possessing little political leverage himself, the lumber worker was the beneficiary to some extent of the efforts of organized labor throughout the state and the nation.[28]

For years it had been the policy of many employers to require workers to sign contracts relieving the employer from liability in case of accidents. Some courts held such contracts to

be against public policy and thus null and void, and before 1908 a number of states, including Texas, took legislative action to outlaw the practice.[29] It had also been commonly ruled that the right of action against an employer expired with the death of the injured worker. This rule would relieve the company of all responsibility in a fatal accident and deprive the survivors of any recourse. Texas, along with thirty-eight other states, abolished this old common-law rule before 1904.[30] The traditional common-law defenses against liability were also under attack. By 1908 the Texas Federation of Labor had lobbied successfully for the abolition of the rule concerning the "negligence of a fellow-servant" as applied to railroad workers and the restriction of its application in other industries. The use of the doctrines of "assumption of risk" and "contributary negligence" was also restricted and modified.[31]

Such measures set the stage for the enactment of a system of workmen's compensation. Such legislation, based on the proposition that the cost of industrial injuries should be borne by industry regardless of who was negligent, had been enacted in western Europe in the late nineteenth century. By 1910 most western European countries, including Germany, Great Britain, and France, had adopted some system of workmen's compensation providing for a schedule of payments to the worker in case of injury. In the United States similar legislation met with adverse court rulings holding that it deprived the parties of trial by jury and the employer of his property without due process of law. Following a period of legislative investigations, during which the need for a more efficient system to replace employers' liability suits was demonstrated, several of the more

---

[28] Brandeis, *History of Labor in the United States*, 3:517; M. W. Splawn, "A Review of Minimum Wage Theory and Practice with Special Reference to Texas," *Southwestern Political Science Quarterly* 1 (March, 1921):4.

[29] Brandeis, *History of Labor in the United States*, 3:566.
[30] Ibid.
[31] Ibid., p. 569. Texas did not go as far in this period (1900–12) as did some states, such as Wisconsin, which outlawed all the traditional common-law defenses. See Robert S. Maxwell, *La Follette and the Rise of the Progressives in Wisconsin*, pp. 153–58.

progressive states passed legislation designed to protect the worker and at the same time meet the criticisms that such laws had previously received in the courts. Wisconsin passed the first so-called modern workmen's compensation law, which won approval of the courts in 1911. The technique employed was to pass a "voluntary compensation law" that in reality coerced the employer to come under it. Within the next two years about twenty other states, including Texas, passed and put into operation similar workmen's compensation laws.[32]

In Texas agitation for a workmen's compensation law had grown steadily since the turn of the century. The Texas Bureau of Labor Statistics announced that in 1908–1909, among 23,253 employees in all industries, 2,820 accidents had been reported, 34 of which were fatal. The Texas State Federation of Labor demanded a comprehensive system of compensation that would automatically cover all workers. In 1911 the Texas legislature appointed a commission to study the question, and, although the commission failed to issue a report, considerable information was assembled and passed on to the lawmakers. Experts in the field estimated that under common law 80 percent of injuries and deaths were denied relief. The Texas state Democratic platform of 1912 included a pledge for a compensation law: "We favor the enactment of an employees compensation law affording adequate indemnity for injury to body or loss of life applicable to employees in this state engaged in hazardous avocations." With Governor O. B. Colquit on record in favor of this legislation and with support from both employees and employers, the 1913 legislature promptly passed the first state workmen's compensation law. It was, as one writer observed, "the response of a progressive legislature to a national movement."[33]

The law of 1913 by no means ushered in the millennium for the sawmill worker, but it was a great step forward. The legislation set up the Industrial Accident Board, to which was entrusted general administrative responsibility for putting the act into operation. Employers were organized under the Texas Employers' Insurance Association. An employer with three or more employees could decide to become a "subscriber" or not, as he chose. The law, however, contained incentives to coerce him to join. For example, if he subscribed to the association, an injured employee was barred from suing for damages, except in cases of willful intent to injure. On the other hand, if the employer chose not to join the association, the law denied the employer the use of the three traditional common-law defenses. The employee might also elect not to be covered by the compensation system, but in actual practice that proved to be of no major consequence. The plan started modestly, setting a maximum of 60 percent of the injured person's weekly wage as compensation and prescribing a seven-day waiting period. Most employers elected to come under the system, chose a responsible insurance representative, and regularly filed reports with the Industrial Accident Board.[34]

Under workmen's compensation the employee automatically received payments for time lost through injury on the job. A typical example was that of an employee of the Angelina County Lumber Company, a $2.50-a-day laborer, who was injured while working in the mill. According to the accident report, "A

---

[32] New York had passed such a law in 1910, but it had been declared unconstitutional by the state supreme court. Kansas and Washington also passed compensation laws early in 1911, but the Wisconsin statute had gone into operation earlier. Maxwell, *La Follette and the Rise of the Progressives in Wisconsin*, pp. 157–60; Brandeis, *History of Labor in the United States*, 3:570–75.

[33] Charles E. Clark, "The Texas Workmen's Compensation Act," *Vernon's Annotated Civil Statutes*, 22:xvii–xlvii; Ernest W. Winkler, *Platforms of Political Parties in Texas*, University of Texas Bulletin, no. 53 (1916), p. 584.

[34] Clark, "The Texas Workmen's Compensation Act." Among the major lumber manufacturers, to the authors' knowledge, only W. T. Carter and Brother Lumber Company declined to become affiliated with the workmen's-compensation system but set up its own system of self-insurance and compensation.

stick was hung under a slasher saw and he had another stick in his hand trying to push the stick out from under the saw, his foot slipped and he fell, sticking little finger of right hand in saw cutting same considerable." The accident board ordered compensation of $21.76, somewhat less than 60 percent of his lost wages for sixteen of the twenty-three days that he was off work.[35]

In 1917, under the leadership of Governor James E. Ferguson, the law was revised and extended; in 1923 the schedule of payments was increased; and in 1927 the seven-day waiting period was eliminated. These changes brought the Texas Workmen's Compensation Law more in line with the laws of the leading industrial states.[36]

The related question of safety in the lumbering industry became more acute during the same years. The absence of guards from saws and cutting tools, the proximity of whirling belts that might break, and the constant danger of power-driven conveyers that might entangle a man's clothing or arm attracted increasing attention among safety experts and others. With the passage of the state workmen's compensation acts the insurance companies involved began bringing greater pressure on insured companies to install safety devices. As the costs of industrial accidents became more apparent to lumber executives, they began taking stronger measures in accident prevention. Safety devices, however, were often expensive and were added slowly. At the end of the 1920s square cutting heads on planing machines were still being operated alongside machines having the safer rounded or circular cutting heads. Such machinery often was not replaced until it wore out. Despite the greater emphasis on safety that accompanied progressive labor legislation, lumbering re-mained a most hazardous occupation, and the state of Texas lagged behind many other states in enforcing such laws as were on the books.[37]

The lumber workers of Texas found an ally and champion in the person of Dr. Benjamin F. Bean, of Kirbyville. Dr. Bean, a medical doctor and politician, was strongly anti–lumber baron and, seemingly, especially anti-Kirby. The son of a Georgia Baptist minister, Bean came to Texas as a young man and served five terms in the Texas legislature. From 1899 to 1905 and again from 1925 to 1927, Bean was the representative from Jasper and Newton counties. In all his legislative activities he consistently worked in behalf of the workingman and against big business, including lumber, railroad, and oil companies.[38]

Bean introduced a number of measures designed to aid the lumber worker or curb the power of the lumber barons. In 1899 he promoted the passage of an act recognizing the right of the workers to organize, strike, and picket—but not trespass. In 1903 he sponsored an anticoercion bill, protecting the worker from blacklisting and from reprisals for not trading at the company commissary. In the same legislature he offered a bill prohibiting compulsory medical deductions, but in the face of strong opposition from millowners the measure failed.[39]

Bean's chief target was the merchandise check in lieu of wages. In this battle he had many allies, including the Texas State Federa-

---

[35] "Report of an Accident to an Employee," Industrial Accident Board, State of Texas, May 8, 1925, Angelina County Lumber Company, ACLCo. Papers.

[36] Clark, "The Texas Workmen's Compensation Act"; Allen, *East Texas Lumber Workers*, p. 109.

[37] Clifton McCleskey, E. Larry Dickens, and Allan K. Butcher, *The Government and Politics of Texas*, pp. 440–41; T. H. Mastin to Angelina County Lumber Company, August 2 and 22, 1925, ACLCo. Papers; Texas Commissioner of Labor Statistics, *Biennial Report, 1913*, quoted in Allen, *East Texas Lumber Workers*, pp. 110–11.

[38] Conversations with Ferris D. Bean, Kirbyville, August 4, 1964; *Members of the Legislature of the State of Texas from 1846 to 1939* (August, 1939), p. 140; conversation of Maud Bean Wetzel with Harry Weaver, June 30, 1960, quoted in Weaver, "Labor Practices in the East Texas Lumber Industry to 1930," p. 55.

[39] H. P. N. Gammel, *Laws of Texas, Supplement*, 1:89, 262; Eli Wiener to W. B. O'Quinnto, January 30, 1903; O'Quinnto to Wiener, February 3, 1903; ACLCo. Papers.

tion of Labor. As early as 1897, during Bean's
first term as state representative, he supported
an act directing that "all wages should be due
and payable weekly or monthly, and should
be made in the lawful money of the United
States." Exceptions and loopholes rendered
the measure ineffective, and in 1901 Bean promoted a bill declaring it illegal for any company to "issue any check or token obligatory to
any employee redeemable in goods or merchandise." Owing to lobbyists' pressure this
bill failed also. Finally, in 1905, Bean was instrumental in securing passage of a law with
teeth, which prohibited the issue of all merchandise checks or tokens and made violation
a criminal offense punishable by fine or im-
prisonment. This measure, which promised to
be both effective and stringent, was promptly
challenged in the courts. After two years of litigation the Texas Supreme Court declared, in
Jordon v. *Texas,* that the 1905 law was unconstitutional. Following the reasoning of the decision in *Lochner* v. *New York,* the presiding
justice denounced the act as depriving the
worker of his freedom of contract and the
company of its property without due process
of law.[40]

After that session Bean retired from the legislature and except for a single term in 1925–
27 devoted himself to his medical practice and
to various business activities in Kirbyville,
where he could be depended upon to oppose
the desires and wishes of the Kirby Lumber
Company and John Henry Kirby. In 1914,
Bean supported James E. Ferguson for governor and strongly applauded when "Farmer
Jim" secured the passage of an act requiring
all manufacturing companies to pay twice a
month in "lawful money of the United States."
Like the earlier legislation this measure lacked
stringent penalties for violation and had no
effective means of enforcement, and so was
largely ignored. The opposition of Bean and
the nearby town of Jasper so angered the
"Prince of Pines" that, according to local tradi-

tion, he ran his railroad (later a branch of the
Atchison, Topeka and Santa Fe) a mile east of
Jasper. Bean made many enemies, but he also
had loyal friends and strong supporters. His
efforts in behalf of the lumber workers probably helped institute the labor laws of the
1930s.[41]

Texas lumber workers were unorganized
and remained unorganized throughout the
fifty-year period of the bonanza era. Indeed,
union activity was infrequent and without exception unsuccessful. That is not to say that
the workers were contented and complacent,
for often they were not. But the complaints
and walkouts were unorganized, sporadic, and
local. When a specific issue was resolved, the
workers were uninterested in attempting any
permanent organization. The Knights of Labor, the American Federation of Labor, and
the Brotherhood of Timber Workers were unable to gain a foothold in the East Texas
pineries.

The details of labor walkouts and strikes illustrate how unorganized and uncoordinated
these uprisings were. In August, 1877, workers
in the Harrisburg sawmills rioted and threatened to burn the town. In 1886 (the year of the
Knights of Labor strike against the Texas and
Pacific Railroad) the workers at a sawmill at
Eylan went on a successful strike for a shorter
workday. In the 1890s there was periodic unrest in Orange over hours and working conditions. Unsuccessful strikes took place there in
1890, 1893, and 1894, but evidently the workers eventually won, for in 1901 Lutcher and
Moore established a ten-hour day, which was
soon copied by the other mills in the Orange
area.[42]

Sporadic strikes continued among Texas
sawmill workers during the next decade, but

---

[40] *Ed Jordon v. Texas,* 51 *Tex. Crim.* 531 (1907).

[41] Conversation with Ferris D. Bean, Kirbyville, August 4,
1964; conversation of Maud Bean Wetzel with Harry Weaver,
June 30, 1960, quoted in Weaver, "Labor Practices in the East
Texas Lumber Industry," pp. 55–56.

[42] *Galveston Daily News,* August 2, 7, 1877; December
14–16, 1890; Lutcher and Moore Papers, Home Office, Orange; Allen, *East Texas Lumber Workers,* pp. 165–67.

they gave little indication of union organization and leadership. The workers at the Kirby mill in Beaumont went on strike in October, 1903, because a promised payroll failed to arrive. The Trinity County Lumber Company faced a strike among logging workers in 1904, and a strike and lockout occurred at a small independent mill at Kirbyville. In 1907–1908, owing to the so-called Bankers' Panic, many mills announced retrenchment policies that reduced wages, shortened the work week, and laid off many workers. The reaction was a flurry of abortive strikes, but with the industry in temporary depression and hundreds of men looking for work, these were foredoomed failures. Within a few weeks the workers were back at work, on the company's terms.[43]

About 1910 the Brotherhood of Timber Workers (BTW) moved into Louisiana and East Texas and tried to organize the logging and sawmill workers. The announced goals of the brotherhood as listed in the *Rebel*, its official publication, were as follows:

1. A minimum wage of $2.00 per day, the work day not to exceed ten hours in duration.

2. A two weeks pay-day in United States, not commissary, currency.

3. The right of free trade, the workers not to be forced to buy from the Company Stores, where prices are 33 1/3 per cent to 50 percent higher than in surrounding "free towns."

4. A discontinuance of the practice of discounting wages.

5. A reasonable rent.

6. A revision of insurance, hospital and doctors fees, the men to have the right to elect their doctors, to see the insurance policy, and to have representatives on a committee that is to control these funds.

7. A general improvement in the sanitary and living conditions of the lumber towns and camps.

8. The disarming and discharge of all gunmen (company guards).

9. The right of free speech, press, and assembly.[44]

These demands, which remained virtually unchanged throughout the entire active career of the brotherhood in the Southwest, appeared reasonable and moderate considering the many ills that the lumber worker endured. They appalled the operators, however, who closed ranks to oppose and destroy the brotherhood, which they denounced as alien, socialistic, wild, and lacking in sanity. The editor of the *St. Louis Lumberman*, a leading trade paper, pledged that the manufacturers would "oppose, with every resource at their command, the introduction of the proposed labor system into the plants."[45] Perhaps the owners' greatest objection was the prospect of having to deal with a union and its outside representatives. They announced that, rather than submit to such demands, they would order "wholesale shut-downs" of their plants.[46]

In the summer of 1910 the BTW led a strike against the Sabine Tram Company at Deweyville aimed at a shorter workday. After about two months the strike failed, the mill reopened, and the workers who were union members were discharged. Other mills followed the same pattern of opposition: refusing the demands of the union, shutting down when necessary, importing scabs (nonunion men) to replace the strikers, and weeding out the labor organizers and union members. At a meeting of the Sawmill Operators Association in New Orleans in 1911 the members gave the executive committee authority to order the closing of about three hundred mills in Louisiana, Texas, and Arkansas, as it thought necessary.[47]

[43] *Orange Leader*, October 16, 1904; Texas State Federation of Labor, *Proceedings*, *Seventh Annual Convention*, 1904, pp. 30–32; Allen, *East Texas Lumber Workers*, pp. 167–72. Among the companies that faced strikes in the Panic of 1907 were the Kirby mills, the Lufkin Land and Lumber Company, the Southern Pine Lumber Company, the Orange Sawmill Company, and the Central Coal and Coke Company.

[44] The *Rebel*, February 8, 1913, quoted in Charles R. McCord, "A Brief History of the Brotherhood of Timber Workers" (master's thesis, University of Texas, Austin, 1959), p. 38. Almost identical demands are quoted from the *Industrial Worker*, December 26, 1912, in Selig Perlman and Philip Taft, *History of Labor in the United States*, *1896–1932*, 4 vols. (New York, 1935), 4:245.

[45] *St. Louis Lumberman*, August 1, 1911, quoted in McCord, "A Brief History of the Brotherhood of Timber Workers," p. 39.

[46] Ibid.

[47] Allen, *East Texas Lumber Workers*, p. 180.

Although the brotherhood's organizing drive caused some labor activity unrest in East Texas, most of the action took place in Louisiana. At these bitter, intermittently violent strikes and lockouts most East Texas owners and operators were concerned and nervous spectators. Yet the manufacturers in both states were closely allied and united through both the Sawmill Operators Association and financial interests. Such owners as Lutcher and Moore, Kirby, Pickering, Frost, and Long-Bell operated lumber mills in Louisiana as well as in Texas. Consequently all the southwestern lumber barons had a personal stake in the outcome of the Louisiana disputes.[48]

The union's efforts to organize the workers in 1911 were largely unsuccessful. Employers anticipated union demands by resorting to lockouts and required workers to sign cards declaring that they would not join the BTW. Such strikes as the union did call failed, and the brotherhood members, estimated at between five thousand and seven thousand in Texas and Louisiana, were blacklisted. When the brotherhood began considering an affiliation with the Industrial Workers of the World (IWW), the Southern Lumber Operators' Association devised a form on which the worker himself asked that information about his work record be sent from his former employer to a prospective one. A question frequently asked about a prospective worker was, "Have you any reason to believe that he sympathizes with or is a member of the order of Timber Workers of the World?" By the winter of 1911 the mill owners were confident that they had destroyed the brotherhood or at least neutralized it, and virtually all mills had reopened and were operating full time.[49]

The BTW set up regional headquarters in Alexandria, Louisiana, under the leadership of President A. L. Emerson. There the brother-

hood held its second annual convention in early May, 1912, and approved the proposed affiliation with the IWW. Both Bill Haywood and Covington Hall (national leaders of the IWW) were present and argued the case for the merger. Despite protests from the local police, the convention seated the white and black delegates together, arguing that the state had not objected when these men worked side by side and should have no complaint when they met to deliberate together.[50]

The formal affiliation of the Brotherhood of Timber Workers with the IWW gave the owners additional ammunition against the union organizers. The IWW had the reputation of being extremely radical, given to violence, Marxist in philosophy, antisegregationist in its organization, antireligious, and committed to class war. It is not surprising that these characteristics, real or alleged, alienated most Texas workers from the BTW and its new parent union.[51]

Within a few days after the convention adjourned, the BTW presented a series of demands to the mills in the De Ridder, Louisiana, area. When the demands were rejected, the union went on strike and set up pickets. The operators' association responded by enforcing a lockout and firing all workers suspected of sympathy with the brotherhood. Actually, the year 1912 should have been more favorable for union activity than the summer of 1911. In 1911 the lumber business was in a slump, prices were said to be down, and most companies had large inventories on hand in their yards. By the spring of 1912, however, business was better, prices were rising, and the large inventories had been depleted. Thus the owners stood to lose money in any prolonged shutdown, and they actively sought nonunion workers, including farmers, to reopen the

[48] See letters on this topic in the ACLCo. Papers, especially Southern Lumber Operators' Association to Angelina County Lumber Company, April 15, 1912, ACLCo. Papers.

[49] McCord, "A Brief History of the Brotherhood of Timber Workers," pp. 31–36.

[50] Ibid., pp. 50–53; Perlman and Taft, *History of Labor in the United States*, 4:245.

[51] Ruth Allen, professor of economics in the University of Texas, after studying the record, concluded that "the merger was a mistake." The BTW was, however, in need of funds and outside support in 1912, and only the IWW stood ready to help. See Allen, *East Texas Lumber Workers*, p. 183.

Standard laborer's contract, 1912. *Courtesy of Stephen F. Austin State University, Forest History Collections.*

To   S. W. HENDERSON, V.P. & GEN. MGR.,

KELTYS, TEXAS.

A. & N. R. RY. CO. _____ *Company.*

Having been employed by said Company from ____ Aug 1908 _____
<small>(Month and Year)</small>

to _____ Jan. 1912. _____ as _____ Tel. Opr. & Conductor. ____
<small>(Month and Year)</small>   <small>(Position)</small>

at ____ Keltys, Texas. _____ under _____ Your supervision. ____
<small>(Location)</small>   <small>(Name of Superintendent or Manager)</small>

of _____ I beg to request that you furnish
<small>(Address)</small>

Mr. Chas. V. W. Schmidt, Superintendent, Railroad Department, the United States Fidelity and Guaranty Company, Baltimore, Maryland, a full transcript of your record of my service as such employe, hereby agreeing that the same is furnished at my express request and for my benefit, and expressly covenanting that nothing contained in the said record (if anything) derogatory to my character or service, nor the furnishing of the transcript of the record, shall be in anywise excepted to or complained of by me, and that the said Company shall be in nowise answerable or liable to me on account of anything therein contained.

WITNESS my hand and seal this _10th_ day of _May_

A. D. _1912_

Witness :

Records transfer request, from the Angelina County Lumber Company papers. *Courtesy of Stephen F. Austin State University, Forest History Collections.*

mills. To counter this move, the union called a series of mass meetings at which the leaders delivered speeches and passed out literature urging all workers to support the union and rally to promote solidarity. They called on the workers not to scab, but the nonstrikers were regularly protected by armed guards, and the union could use only persuasion.[52]

On July 7, 1912, a mass meeting was held at Graybow (or Greybeau), Louisiana, before the offices of the Galloway Lumber Company, and the Union president, Emerson, tried to speak. Almost immediately shooting commenced, and the crowd, numbering about 150, screamed and sought cover. Eyewitnesses estimated that about three hundred rounds were fired, the shots coming from both the company offices and union members in the crowd. Three men were killed, a fourth later died of his wounds, and more than forty were wounded. Three of those who died were union men, and one was a company guard. The press described the incident as a "riot" and charged that the union men had come to shoot up and take over the town. Emerson spoke of it as a massacre, a deliberate attempt to "shoot us down and murder me." The sheriff of Calcasieu Parish and his deputies arrested sixty-two union men, including Emerson, and charged them with murder. A grand jury duly indicted fifty-eight of them and dropped murder and assault charges against the company officials and guards.[53]

The trial of a selected nine of the defendants, including Emerson, took place at Lakes Charles in October, 1912, and was reported throughout the country. After long delays, during which all fifty-eight men languished in the parish jail, the spectacular case got under way, and after much conflicting testimony the case went to the jury. After only an hour or so of deliberation the jurors found the defendants not guilty. Soon afterward all fifty-eight were released.[54]

Although Emerson proclaimed the verdict as a great victory for labor, the trial virtually bankrupted the BTW, frightened many lumber workers, and largely discredited the union in the eyes of the public. Membership, which had been estimated at between 18,000 and 35,000 in Texas, Louisiana, and Arkansas, began to decline. Following an unsuccessful strike at Merryville, Louisiana, against the American Lumber Company, in which company guards sacked the union headquarters and attacked President Emerson, the disintegration of the BTW in the South proceeded rapidly, and by the end of 1914 it was virtually extinct, defeated by the superior economic power and organization of the employers.[55] It was not until the New Deal period that unionism made any headway among the lumber workers of the Gulf Southwest.[56]

Texas lumber workers made wage gains during World War I but not as many as those of West Coast lumber workers or Gulf Coast refinery employees. Between 1916 and 1919 the hourly wages of laborers in Texas sawmills doubled, from 15.5 to 31 cents an hour; skilled workers' hourly wages also doubled, from 24 to 47 cents. In comparison, the wages of the common laborer in Oregon rose from 20 to 52 cents an hour; the skilled worker's wages increased from 35 to 80 cents an hour. In Texas oil refineries by 1919 a common laborer was receiving about 40 cents an hour, 8 or 9 cents more than his counterpart in the sawmill.[57]

As might be expected, the industry did what it could to discourage migration of lumber workers to the Gulf Coast or to other parts of

[52] McCord, "A Brief History of the Brotherhood of Timber Workers," pp. 53–55.

[53] Ibid., pp. 56–64.

[54] McCord goes into the evidence of the trial in considerable detail, securing information from a number of sources. A point

that evidently weighed heavily with the jury was that apparently the first shots came from the company offices. McCord, "A Brief History of the Brotherhood of Timber Workers," pp. 74–93.

[55] Ibid., pp. 94–106. Technically the BTW continued to exist until 1916, but by then its most loyal members knew that it was dead.

[56] Perlman and Taft, *History of Labor in the United States*, 4:247.

[57] U.S. Department of Labor, Bureau of Labor Statistics, *Industrial Survey in Selected Industries in the United States, 1919*, Bureau of Labor Statistics Bulletin no. 265 (1920), p. 348. U.S. Department of Labor, Bureau of Labor Statistics, *Wages and Hours of Labor, 1916*, pp. 234–36.

the country in pursuit of better wages. Nevertheless, there was a marked migration of blacks from Texas during the war years, which left a "great scarcity" of common labor. So serious was the shortage that the Southern Pine Association offered, under certain conditions, to advance transportation costs to groups of blacks who wished to return to a particular mill. The money was to be refunded later by the employees. The association blamed the United States Department of Labor for recruiting labor in the South during the war years and encouraging black sawmill hands to migrate North, only to find themselves out of work when the emergency was over.[58]

Although during the 1920s wages did not return to the prewar levels, they did drop some owing to the declining state of the lumber industry. A Texas state survey in 1927 disclosed that unskilled laborers received an average of $2.59 a day and were increasingly irregularly employed. Federal reports listed Texas common labor at $.24 an hour with average weekly earnings of $11.57. In comparison West Coast workers were paid more than $.30 an hour, and most made more than $20.00 a week.[59] It became increasingly difficult for a lumber worker, even a skilled artisan such as a sawyer or an edgerman, to change employment or find a new job if his mill closed. It was not unusual for experienced sawyers to take lesser jobs simply because there were no openings for sawyers in the entire East Texas area. The surplus of workers in all categories tended to depress wages and precluded a further reduction of hours. Well before the onset of the Depression in 1929, the Texas-Louisiana lumber industry was in a depressed condition, and workers were seeking employment on almost any terms. As more and more mills closed, production declined, and the industry all but collapsed. Saw-

mill employees turned to other industries or farming or went on federal relief (after 1933).[60]

Thus ended the bonanza era in Texas. For fifty years the lumber industry had provided employment for several thousand unskilled, semiskilled, and skilled workers. Wages were low, hours were long, and the work was hard and often dangerous. There were opportunities for advancement, and some young men made their way up through the ranks to positions of responsibility and authority. But the great majority of unskilled workers remained unskilled workers and black workers found that the way up was even more restricted and limited. The lumber worker made little progress in unionization or in improving his bargaining position with his employers. During the entire half century he was able to increase his basic wage only $1.00 a day, from $1.50 to $2.50, and to reduce the length of his workday from eleven to ten hours in a six-day week. In comparison with the industrial worker or the West Coast lumber worker he was in a worse position in the 1920s than he had been in 1890.

Wages and hours were not the whole story, however. The logger and the sawmill worker of the bonanza era lived in a subsistence economy and a company-oriented world. Most workers' circle of experience was restricted to other lumber companies and the thin, poor farms of East Texas, where life was hard and prospects were even less promising than those at the mills. Most workers were content to "get along," to rise slowly in the hierarchy of lesser managerial positions, and to trust that the company would help them if they were laid up by injury or illness. Indeed, the company filled most of the workers' lives. Most of their stories, jokes, and anecdotes concerned the owners, the bosses, or fellow workers. Loggers and

[58] Allen, *East Texas Lumber Workers*, pp. 56–57.

[59] Texas Bureau of Labor Statistics, *Biennial Report, 1927–28*, pp. 44–45, 120–25; U.S. Department of Labor, Bureau of Labor Statistics, *Wages and Hours of Labor, 1929*, p. 35.

[60] Conversation with Dr. George F. Middlebrook, Caro, October 15, 1959; conversation with Clyde J. Woodward, Sr., Nacogdoches, October 3, 1957.

mill men regaled their families with the happenings of the day—who said this, and who was gigged for that. Through all their labors there was evident a strong, professional pride in their work, a kind of exultation that they were part of a demanding and dangerous industry. The great majority of workers liked their jobs. They were loggers or sawmill men by choice, and they would not willingly exchange their life for any other.[61]

[61] Conversation with Roy Dudley, Groveton, March 15, 1960; conversation with W. S. Brame, Livingston, 1959; conversation with Charlie Welch, Brookland, August 5, 1958; conversation with George Morrison, Huntington, August 8, 1958; conversation with Needham B. Weatherford, Camden, October 13, 1959; conversation of John Larson with D. D. Devereaux, Diboll, 1954, Forest History Society, Santa Cruz, Calif.

CHAPTER 9

Company Town, Company Store

In the Feudal Age, the villages of the re-
tainers huddled at the foot of the overlord's
castle. In Twentieth Century America the
feudal communities cluster around . . . the
sawmills.—George Creel

ONE characteristic feature of the Texas lumber industry in the bonanza era was the company town, which was completely dependent on the sawmill and its owners for survival. The lumber company town was not a phenomenon in American history. In the Appalachian Mountains coal-mining towns thrived for more than a hundred years. The gold-mining towns of the West, such as Virginia City and Deadwood, were essentially company towns. In New England in the nineteenth century and the southeastern states in the twentieth, company towns clustered around the textile mills that were the sole employers.[1]

The East Texas lumbering company town was similar to, yet distinctive from, the company towns of other industries. It was generally isolated, far from any center of population, in a primitive section of the state. It was usually served by a single railroad, which often was also the property of the sawmill company. The only other routes of travel were unimproved roads, often little better than trails, and most lumber workers did not possess any means of transportation except their two feet. Most often the company owned the entire town—land, streets, wells, houses, stores, churches, schools, lodge halls, and any other

buildings. It produced a peculiarly dependent situation for the residents.[2]

The reasons for the development of the Texas sawmill towns are clear. The owners built their first sawmills in the Piney Woods at or near sources of timber. It was much more economical to transport finished lumber to retailers than to haul logs great distances to a mill. The construction of the mill created job opportunities for hundreds of workers where neither workers nor jobs had existed before. The owner had to provide living quarters for his employees, housing usually being included in the pay. Unlike his Great Lakes states or Canadian counterparts, the Texas logger or sawmill worker was usually a married man with a family, who accompanied him to his new job. Thus began the East Texas company town in the 1880s and 1890s.[3]

The ethnic background of the towns was remarkably homogeneous. Except for the black population, which was as segregated in the milltowns as it was elsewhere in the South, the inhabitants were overwhelmingly Anglo-Saxon, native-born, southern, white, and Prot-

[1] For company towns in American industry see Carroll R. Daugherty, *Labor Problems in American Industry*, pp. 279, 614–17, 654–57.

[2] Conversation with Clyde J. Woodward, Sr., Nacogdoches, October 3, 1957; Oral History Collections, Forest History Collections, Stephen F. Austin State University Library, Nacogdoches; hereafter cited as SFA Library. Unless otherwise noted, interviews and conversations cited are in these collections, and all interviews were conducted by principal author Maxwell.

[3] Conversation with Needham B. Weatherford, Camden, July 25, 1958.

A better-than-average company house, probably occupied by middle management. *Courtesy of Ste-* *phen F. Austin State University, Forest History Collections, Thompson photo albums.*

estant, chiefly evangelical-fundamentalist. In these characteristics the people of the sawmill towns did not differ substantially from other residents of the Piney Woods. Nor did they differ from other East Texans in their level of education. The census returns of 1900 showed that in most of the East Texas lumbering counties the proportion of adult males classed as illiterate regularly ran higher than 20 percent, in comparison to an illiteracy rate of 15.4 percent for the state.[4] One can assume that the wives' educational level was comparable.

A typical, if somewhat more attractive than average company town was Camden, the

headquarters of the W. T. Carter and Brother Lumber Company. According to veteran workers, when the town was built in 1898, no definite plans were made for development and growth. The streets were not laid out in blocks or squares; the principal streets curved around a hill, two streets forming a complete circle, commonly called the "Loop." The streets were unimproved and were largely sand. By 1914 about 450 houses had been built for the employees. A row of houses for the managerial personnel and Carter's house, standing somewhat apart on the hill, were also completed about the same time. In the business section of town were the general store or commissary, the hotel—boarding house, the company offices, and the depot for the Moscow, Camden and San Augustine Railroad. Water was supplied

[4]U.S. Department of Commerce, Bureau of the Census, *United States Census, 1900,* vol. 1, table 92, quoted in Ruth A. Allen, *East Texas Lumber Workers: An Economic and Social Picture, 1870–1950,* p. 67.

Company house. *Courtesy of Stephen F. Austin State University, Forest History Collections,* *Thompson photo albums.*

by wells, each serving one to four houses; before 1926 none of the employees' houses had running water.[5] Water was hand-carried, usually by the children, from the well to the kitchen. Sanitary facilities were simple, and baths were infrequent. The kerosene lamp was the standard light in the home until electricity became available in 1926. Because the workday was long, the labor hard, and reading minimal, it is probable that the lamps were used sparingly.

The employee's house was typically box-shaped, with four rooms and a porch front and rear. It was built of twelve-inch box siding, with a metal or shingle roof, a single floor, and rather high ceilings and stood two to three feet above ground level. Houses of this early period

were seldom painted but were usually kept neat and in good repair. With most homes came garden plots, where the wives raised vegetables and kept a few chickens; some families kept cows in common pastures. Many early residents recalled that livestock wandered freely in the streets and yards. The houses were not screened and were heated by wood-burning cookstoves in the kitchens and pot-bellied wood stoves in the living rooms.[6] The standard of living did not differ markedly from that of most farmers in the Piney Woods region.

No doubt East Texas company towns resembled each other in most aspects, but some were markedly superior or inferior to the aver-

[5] Conversation of Carter S. Caton with Nora Havard, June 14, 1965.

[6] Conversation of Jay Dee Sharp with William W. Kirkland, June 15, 1962; conversation of Kathleen White with Charlotte King Smith, June 26, 1965.

age. Fostoria, a milltown owned by the Foster Lumber Company of Kansas City, was described as probably the "cleanest and nicest mill town in Texas." The houses were painted and were neatly laid out in squares. Situated on the Santa Fe Railroad, the town was self-sustaining, with its own stores, post office, hotel, depot, utilities, schools, athletic field, churches, and amusement center. In contrast, one investigator described Carmona, the Saner-Ragley Lumber Company town, as composed of unpainted houses with sandy streets, minimal cultural activities, and employees frequently on the move. Journalist George Creel described Kirbyville as "gray, dingy boxes ranged row by row in the horror of dull uniformity that is the curse of most industrial communities." Another Kirby town, Browndell, was described as a collection of "rotten shacks, rotten commissary, rotten doctors, rotten insurance." About 1930 the company designated the houses in Keltys as A, B, C, and D types. Type A and B houses, numbering about fifty, had porches, garages, natural gas, sewer lines, electricity, and water. Second and third-echelon management families occupied these houses. The type C and D houses lacked porches and water but did have electricity if the employee installed it. The latter houses were unpainted and rented for eight to ten dollars a month.[7]

A standard feature of the milltown was the combination hotel and boarding house for unmarried workers. It usually stood near the center of the business section—the railway depot, the company offices, the commissary, and the post office. No doubt such establishments varied as widely as the people who operated them. In some towns, such as Camden, the boarding-house manager prided herself on her "good

table," which consisted of plain but well-cooked and attractively served food. The typical meal, served at a long, family-style table, was a generous helping of meat, three or four vegetables, corn bread, coffee, and pie. In other company towns the boarding -house fare must have been unpalatable. Stories about the boarding-house food reached the workers by the grapevine, and they knew which ones to patronize and which to avoid.

Stories and practical jokes about the boarding-house food abounded. In a Kirby town on the Santa Fe a rather fastidious Scot who worked in the local Kirby office was constantly concerned about the nature of the unidentified meat that the cook regularly served in the stew. Knowing this, boarders delighted in joking about the food. Once, according to a veteran lumberman, the housekeeper complained that her dog had disappeared and could not be found. The next evening a particularly unappetizing mess was served as "stew." At once the boarders identified the dish as "stewed collie" and opined that it was better than the meat of the previous day, which must have been cat. That was too much for the Scot, who departed abruptly from the dining room with his hands to his mouth. Needless to say "Scotty" was never allowed to forget the "collie stew," and the cook refused to enlighten him about the contents of the dish. At Haslam on one occasion the manager of the mill dumped the contents of a particularly unappetizing plate over the hotel manager's head.

In the years between the turn of the century and World War I boarding-house rates varied from $12.00 to $20.00 a month for board and room. The average seems to have been about $15.00 a month or $0.50 a day. The numbers who lived and ate at the boarding houses varied, but seldom did the manager have more than fifteen or twenty residents, as well as salesmen and visitors. The woods and mill workers had a hearty early breakfast, a cold lunch on the job, and a hot supper in the evening. A few of the men who worked in the company office or commissary regularly ate

[7] Conversation of James Kersh with Hubert Smith, Foster Lumber Company, June 23, 1965; William T. Chambers, "Life in a Southern Sawmill Community," *Journal of Geography* 30 (May, 1931): 181; George Creel, "The Feudal Towns of Texas," *Harper's Weekly* 60 (January 23, 1916): 76–78; Richard G. Lillard, *The Great Forest*, p. 299; Lita M. Mayberry, "Keltys: An East Texas Sawmill Town" (master's thesis, Stephen F. Austin State College, Nacogdoches, 1948).

their noon meals at the boarding house. At some mills the noon whistle signaled a shutdown so that the mill hands could also eat a hot lunch.

In many of the hotels a special suite was kept for the general manager or owner to use when he visited the operation. No doubt the meals improved markedly during the visits of the big boss. R. A. Long, W. A. Pickering, J. Lewis Thompson, and E. A. Frost, among many others, followed the practice of keeping special suites in the local hotel–boarding house for their personal use during periodic visits to their mills. Long on occasion also used a private Pullman for inspection visits.[8]

Most of the millowners provided community or "union" churches that were used in turn by the principal denominations. Thomas L. L. Temple, John Martin Thompson, and R. A. Long were strongly religious men and encouraged Sabbath observance, piety, and sobriety among their workers. Although a few mills ran seven days a week during boom times, this practice was exceptional, and most workers had the traditional seventh day off. Of course, many chose not to attend Church, preferring to fish or hunt, work around the house, sleep, or carouse.

The pattern at Camden was perhaps typical. The company built a union church there soon after the mill was completed. The Methodists, Baptists, and Presbyterians took turns holding services once a month. The Baptists held Sunday school every Sunday morning, and the Methodists every Sunday afternoon. It was said that the churchgoers attended all the services without discrimination. Old-timers admitted, however, that mostly women and children attended church services, especially Sunday school.

Mrs. Frank Farrington, of Diboll, was typical of the religious-minded layperson active in Christian work in the milltowns. According to her own account, T. L. L. Temple encouraged her and her husband to come to Diboll because "there were two saloons and just one little church" in the town. She worked in the commissary, where her husband was manager; taught school; organized a PTA; and promoted Wednesday-night prayer meetings. When the logging-camp workers were moved into Diboll, she was instrumental in organizing a mission church for them. Although she was a Christian Scientist, Mrs. Farrington worked largely through the Methodist church. In addition to promoting Prohibition, she tended the sick and helped the destitute. Many children were outfitted at the commissary at a word from one of the Temples after a solicitation by Mrs. Farrington.

Ministers seldom lived in the milltowns, and most of the church work was carried on by laypersons like Mrs. Farrington. When the minister arrived for the monthly service, he frequently came on Saturday and held two or three services before he departed. Summer revivals were common and were often well attended, perhaps because there were no competing activities. At the end of such a revival it was not unusual for the log pond to be used for a baptismal service.[9]

This pattern of religious activity was followed by most millowners, who encouraged church attendance, provided church buildings, and frequently contributed generously to the visiting minister. A few owners, such as John Henry Kirby, occasionally distributed Bibles to every family on the payroll. The fundamentalist Protestant traditions of most of the workers strongly encouraged worship on the Sabbath. Even the men who found excuses for avoiding

[8] Conversation of Nora Havard with Carter S. Caton; conversation of Nora Havard with Mrs. Jack Watts, June 21, 1965; conversation with Judge J. R. Vaught, Trinity, September 10, 1965; conversation with W. I. Davis, March 20, 1956; conversations with Clyde J. Woodward, Sr., Nacogdoches, October 3, 1957, January 16, 1960; Boarding House Record Book, 1889–90, ACLCo. Papers, Forest History Collections, SFA Library.

[9] Conversation of Kathleen White with Ed Bailey, Pineland, June 28, 1965; conversation of Nora Havard with L. M. Phillips, Camden, June 21, 1965; conversation of Jay Dee Sharp with William W. Kirkland, Lufkin, June 15, 1963; conversation of John Larson with Mrs. Frank Farrington, Diboll, 1965; American Lumberman, September 26, 1908, p. 86.

church attendance urged their wives and children to attend. Perhaps it was the pattern of Christian worship that kept the sawmill community as civilized and law-abiding as it was.

The educational opportunities in the sawmill towns were understandably limited—somewhat more limited than those in the principal towns of the region but somewhat better than those in the surrounding rural regions. In the generation before World War I, Texas lagged behind the rest of the United States in both literacy and years of schooling. Within the state the East Texas literacy level was below the Texas average, and the counties in the longleaf-pine belt had the lowest rate of all. For example, in 1900, when the rate among adult male Texans was 15.4 percent, in Jasper County it was 24.2 percent, in Newton County 18.3 percent, in Sabine County 20.4 percent, and in San Augustine County 28.7 percent. In 1920 the comparative figures showed the rate for Texas as 9.6 percent, Jasper County 12 percent, Newton County 14 percent, Sabine County 13 percent, and San Augustine County 16.9 percent.[10] In years of schooling the census statistics told the same story.

As late as 1940 the census reports showed the East Texas Piney Woods counties lagging far behind the state average and some even below the southern average in total grades completed and in percentage of adult population that had failed to acquire even minimum skills. In East Texas counties the average number of grades completed by adults over age twenty-five ranged from 6.2 to 8.6, with the regional average approximately 7.5 grades. In comparison both the state and the national average approximated 8.5 grades. More than half of the East Texas counties numbered over 25 percent of adults who had completed fewer than five grades (in the census reports five grades being the dividing line between literates and functional illiterates). The state average was 18.8 percent, and the national average was 13.5 percent. The lack of education in the adult population was reflected in the level of educational standards of their children.[11]

The descriptions of the schools in sawmill communities at the turn of the century are very similar. Typically the building, like most of the houses, was a simple square box, largely unfinished on the inside. It was generally divided into two or four rooms, and the number of teachers ranged from one to four, depending on the number of students. Most of the furnishings—benches, desks, and chairs—were homemade, and the blackboards, if any, were crude. Both the school and the teachers lacked state certification, and the school term varied from four to seven months. Seldom did instruction extend beyond the sixth or, occasionally, the eighth grade.

The courses were confined to the fundamentals. Orders for schoolbooks at the Angelina County Lumber Company Commissary included Swenton's *American History*, Monteith's *Geography*, McGuffey's *Revised Speller*, and McGuffey's *Readers*, first, second, third, and fourth. Pens, copybooks, crayons, and paper were also available at the company store.

The school also served as a community center. Both parents and students gathered at the local school for programs including recitations by the students, followed by a spelling bee, in which both adults and children took part, and, finally, refreshments. The school building also served as a meeting hall for discussions of questions of public interest.

In all phases of education the company played a large part. The company owned the buildings and the land on which they stood. The company also paid the teachers, either directly or through taxes paid to the common district on the company's landholdings. Frequently the sawmill company was the only

---

[10] U.S. Department of Commerce, Bureau of the Census, *United States Census, 1900*, vol. 1, table 92; *United States Census, 1910, Supplement for Texas*, table 1; *United States Census, 1920*, vol. 3, table 9.

[11] U.S. Department of Commerce, Bureau of the Census, *United States Census, 1940, Supplement for Texas.*

landowner. In short, education in the sawmill communities was relatively primitive, as it was primitive throughout most of East Texas in this era.[12] Yet occasionally even in such schools as these bright young boys and girls gained a basic education and went on to high school and college or university. The company management frequently loaned or gave funds to help a promising student complete his education.

The company doctor was an important person in the sawmill community. He lived in a comfortable, well-finished house on "managers' row" and was in much demand. His office was usually in his home, but here and there a company built a separate structure for the doctor, a combined office, clinic, and small hospital. The commissary kept a supply of such drugs as the doctor might prescribe.

Not every company town was able to attract or hold a qualified physician. Often the well-meaning wife of one of the officials, such as Mrs. Farrington, ministered to the sick and cared for the injured. Frequently black midwives, such as Aunty Lou Jones and Betsy Black, of Camden, attended expectant mothers, both black and white, and delivered their babies.[13]

The lumber companies normally paid their doctors a salary, funded through hospital fees deducted from the workers' wages each month. The fees were fairly standard throughout the region. The married workers at the Alexander Gilmer Lumber Company in Jasper paid $1.50 a month, and the single men paid $1.00 a month. The married workers at Pineland, the Temple Lumber Company town, paid

$1.00 a month, and the single men $0.75. At Kirbyville the hospital fee was $1.50 for married men and $1.00 for single men.[14] The total fees amounted to considerable sums and no doubt provided the companies with additional profits: there were 150 workers at the Gilmer Lumber Company, 350 at Pineland, and 400 at Kirbyville. The fees were not voluntary or optional; all the workers participated as a condition of employment. This scheme functioned in such a manner that a man who changed employers frequently paid two or, on rare occasions, three hospital fees in a single month.

The hospital fee was one of the chronic complaints of the lumber workers, who considered the charge too high and the services too few and meager. The $1.50 charge represented a day's pay for many of the men, and in the course of the year, they did not average more than twenty-two or twenty-three days' work a month. The fee did not include medicines or coverage for a major illness that might require hospitalization in Houston or Beaumont. Most workers, especially the younger men, would have avoided the fee if it had been optional. Yet perhaps there was no other way that the company could guarantee the doctor a sufficient salary to keep him in an isolated lumber-company town.[15]

The company doctor treated his patients for natural ills and injuries resulting from sawmill or logging accidents. A common ailment was malaria ("the fever"), which was recurrent in its victims. Typhoid and dyphtheria were not uncommon. According to one physician who had served three company towns in turn, the most common illnesses were common colds, influenza, and pneumonia. These and other respiratory diseases were frequent, and tuberculosis was the most feared of the major illnesses. Mill accidents brought to the company

[12] Conversation of Kathleen White with Mrs. L. R. Sanford, Hemphill, June 26, 1965; conversation with Mrs. T. A. White, Hemphill, July 5, 1965; conversation with R. C. McDaniels, Orange, July 4, 1965; conversation of Nora Havard with L. M. Phillips, Sr., Camden, June 21, 1965; *American Lumberman*, September 26, 1908, p. 105; Chambers, "Life in a Southern Sawmill Community," pp. 181–82.

[13] Conversation with Dr. George F. Middlebrook, Caro, October 15, 1959; conversation of John Larson with Mrs. Frank Farrington, Diboll, 1954; conversation of Jay Dee Sharp with Carter S. Caton, June 21, 1963.

[14] Peter A. Speek, "Notes on Investigations of Three Texas Lumber Towns," U.S. Department of Labor, Reports of the Commission on Industrial Relations, 1914 (microfilm), Forest History Collections, SFA Library.

[15] In addition to the hospital fee, for a time after 1913 the companies also collected an insurance fee, usually $.75 to $1.00 a month, under the new Workmen's Compensation Act.

doctor smashed, cut, mangled fingers, hands, arms, and legs. Wood accidents were more serious and caused more fatalities than mill accidents. The rehaul skidder, confirmed the physician, was the source of many logging accidents. Accidents were common in this most hazardous of manufacturing enterprises. The census showed a disproportionately large number of disabled men in the East Texas Piney Woods counties.[16]

In the typical milltown recreation and leisure time were all but nonexistent. The hours were long (ten to eleven hours a day at most mills), the work was hard and exhausting, and recreational facilities were minimal. When business was booming, the mill ran seven days a week for several months. When business was slack, the mill often ran only four or five days a week or shut down for several weeks at a time. Annual production figures indicate that few mills operated more than 260 to 280 days a year, but much of the involuntary leisure time was accounted for by layoffs without pay plus Sundays. Traditionally the company gave a picnic for the white workers on the Fourth of July and a barbecue for the black employees on June 19. (Black Texans celebrate Juneteenth as Emancipation Day, for on June 19, 1865, federal troops arrived in Galveston and announced the end of the Civil War and of slavery.) Most companies also gave a Christmas party or distributed gifts at Yuletide. As mentioned earlier, the largesse of John Henry Kirby at Christmastime was famous.[17]

By far the most popular recreation among the men was hunting and fishing. Not only did these activities cost relatively little, but they brought additional, tasty food to the family larder, and the workers could use the company lands and adjoining streams. The amount of

Thompson and Tucker baseball team, "The Trimmers." *Courtesy of Stephen F. Austin State University, Forest History Collections, Thompson photo albums.*

drinking and gambling varied greatly from town to town. Some owners, such as William Carlisle, of Onalaska; Thomas L. L. Temple, of Diboll; and John Martin Thompson, of Willard, frowned on these activities as vices and took steps to discourage them. Yet there were saloons (before 1918) in most lumber towns, including Keltys, Diboll, Kirbyville, and Nacogdoches. A veteran lumberman recalled that most of the men spent their leisure time sitting around "making axe handles and cussing preachers; played poker and prayed only when they wanted favors."[18]

The pride of many a company town was its baseball team. The teams, made up of the young male employees, played teams from other East Texas towns, usually on Sunday af-

[16] Conversation with Dr. George F. Middlebrook, Caro, October 15, 1959; Allen, *East Texas Lumber Workers*, pp. 113–15.

[17] Chambers, "Life in a Southern Sawmill Community"; conversation of Jay Dee Sharp with William W. Kirkland, Lufkin, June 15, 1963; conversation of Kathleen White with Mrs. L. R. Sanford, Hemphill, June 26, 1965; *American Lumberman*, September 28, 1908.

[18] Conversation of Jay Dee Sharp with William W. Kirkland, Lufkin, June 15, 1963; conversation of John Larson with Mrs. Frank Farrington, Diboll, 1954; conversation with Needham B. Weatherford, Camden, October 13, 1959.

ternoons. It was not unusual for a manager who was a baseball enthusiast to bring in a "ringer"—perhaps a star pitcher from one of the universities—keep him under cover until the day of a big game, and then wager heavily on the outcome. The Thompsons were both supporters of the game and baseball players themselves. They provided an excellent diamond and balls, bats, uniforms, and other equipment for their team, called the Trimmers. Apparently they were one of the outstanding town teams in the Piney Woods in the years just before World War I.[19]

Then there were traveling shows. Occasionally a small circus or animal act came to town with the approval of the millowners, and the workers welcomed this break in the monotony. Patent-medicine shows provided entertainment as well as bombastic oratory that was entertaining in itself. The product was frequently purchased more for its alcoholic content than for its medicinal qualities. One veteran lumberman recalled that "gander dancers" occasionally visited his town and did a dancing act while balancing on each end of a board centered on a sawhorse or a mound of earth.[20]

A few dance halls and roadhouses could be found outside the company town, but the management discouraged the men from visiting them; they were often scenes of violence, especially on Saturday nights. Men who were frequently laid off for minor ailments and stayed in town while the other men were at work were the objects of distrust and fear that they were bent on some adulterous adventure. The owners usually discharged men with such reputations as a matter of policy.[21]

In short, such recreation as there was centered on the church and the school. In a few towns reading clubs and singing societies were organized and flourished for several years. Hunting and fishing were the most popular outdoor activities, and baseball was almost the only organized sport. Politics, religion, and local gossip constituted the chief topics of conversation. It was a hard, simple, narrow life, but a life not markedly different from that in other parts of Texas and the South in that era.[22]

Law enforcement in the company town followed a routine pattern. The local constable performed the usual duties in keeping the peace. There was also a deputy sheriff, whose chief beat was the company town and the company's property. In addition there was often an officer with a commission in the Texas Rangers, locally known as the "quarter boss," who was paid by the company. With his boots, six-shooter, and ten-gallon hat, he had a sobering effect on most wrongdoers, and especially on blacks. It times of tension or labor agitation the ranger was reinforced by deputies, who were especially hated and were described by union organizers as "a bunch of big-hatted boobs who would commit murder or anything else at the command of their chief." The town officials also included a justice of the peace, whose court dispensed justice in the various misdemeanor cases brought before it. A perusal of the criminal docket of the justice of the peace in one company town revealed that the more frequent charges included assault and battery, fighting, gambling and keeping gambling houses, abuse of the Sabbath, and petty theft. Trials for more serious crimes were held in the county court and the district court in the county seat.[23]

[19] Conversation of Jay Dee Sharp with Carter S. Caton, June 21, 1963; conversation with Needham B. Weatherford, Camden, October 13, 1959; conversation with Clyde J. Woodward, Sr., Nacogdoches, October 3, 1957; conversation with W. S. Brame, Livingston, September, 1963; *American Lumberman*, September 28, 1908. The Angelina County Lumber Company baseball team played the Saner-Ragley team in 1891 for a wager of $50 to $25. See Eli Wiener to Will Jennings, Timpson, May 31, 1891, ACLCo. Papers.

[20] Conversation of Jay Dee Sharp with Mrs. W. A. Kirkland, Pollock, June 19, 1963; conversation with Needham B. Weatherford, Camden, June 21, 1963.

[21] Conversation with Needham B. Weatherford, Camden,

October 13, 1959; conversation with Simon W. Henderson, Keltys, August 10, 1964; conversation with Clyde J. Woodward, Sr., Nacogdoches, October 3, 1957.

[22] Allen, *East Texas Lumber Workers*, pp. 67–68; conversation of John Larson with Mrs. Frank Farrington, Diboll, 1954.

[23] Lillard, *The Great Forest*, p. 299; conversation with Clyde J. Woodward, Sr., Nacogdoches, January 16, 1960; conversation of Nora Havard with Byron Pate, Camden, June 14, 1965; Justice of the Peace Criminal Docket, Camden, Polk County, Texas, 1900–48. The puritanical nature of the punish-

Company house in the "quarter." *Courtesy of Stephen F. Austin State University, Forest History Collections, Thompson photo albums.*

In every lumber town, separated from the main area by the log pond, the railroad tracks, or the mill itself, was the "quarter." This was the part of the company town set aside for the black workers in the mill, the yard, the construction gangs, and the logging crew. There they and their families lived, completely segregated as in other southern towns at the turn of the century. The blacks' houses, according to one investigator, were "pretty sorry." They were similar to the smaller houses in the white section except that they were seldom kept in repair. The prevailing style was the unpainted three- or four-room box built of rough lumber with a board or shingle roof. Not only were the houses poorer than those in the white section, but the "quarter" was less well drained and

had few or no trees and few plots for vegetable gardens. Almost no houses had running water, and several houses shared a single water faucet in the yard. Inside the typical house had only a single pine floor, unfinished walls and ceilings, broken windows, and a leaky roof. For this the black worker paid four to six dollars a month, and the rent was often collected weekly.[24]

Blacks made up one-third to one-half of the work force in most mills.[25] Classed almost entirely as unskilled laborers, they had even less money to spend than their white coworkers had. Experiencing the general prejudice against their race, they were usually the last hired and the first laid off, and they lived in constant fear of possible violence at the hands of a white mob. Yet the black sawmill worker

ments meted out is apparent from the records. For example, violating the Sabbath or keeping a gambling house drew much heavier fines than fighting in public, assault and battery, or even attempted rape.

[24] Payroll Books, 1897, Alexander Gilmer Papers, University of Texas, Austin; Speek, "Notes on Investigations of Three Texas Lumber Towns"; conversation of Weldon Hunt with Pope Ware, Sarber, June 15, 1965; Mayberry, "Keltys."

[25] Allen, *East Texas Lumber Workers*, p. 54.

General view of the "quarter." *Courtesy of Stephen F. Austin State University, Forest History Collections, Thompson photo albums.*

was an excellent laborer and essential to the success of the industry.[26]

There were usually both a church and a school in the "quarter," built by the company but after that ignored. The church was most often a one-room boxlike structure with crude benches and little ornamentation. The preacher was usually a layperson who during the week worked at the mill or in the woods. The school was customarily equipped with desks and seats discarded from the white school and with hand-me-down books that had become too worn and torn for use across the millpond. Yet there was regularly a three- or four-month term each winter and instruction to the level of about the sixth grade. Educational opportunities were poor but not markedly poorer than those for black children in small towns and rural regions throughout Texas and the South.[27]

Recreation in the "quarter" was even more limited than in the main section of the town. Certainly, any sports or games that the blacks played they played among themselves. White workers at the managerial level regarded the blacks as more inclined to gamble, drink excessively, and brawl than the whites. One company doctor recalled that it was customary on Saturday nights to be called out to treat stab or slash victims from the "quarter." According to some, the black worker was superstitious, even practicing a variety of voodoo imported from the Louisiana bayous. Certainly charms and good-luck pieces were common.[28]

Yet, although poorly treated and seldom promoted, the blacks were frequently among the workers most loyal to the lumber company and its owners. Their loyalty is illustrated by a speech made in 1940 by Landy Parham, a long-term black employee of the Foster Lumber Company at Fostoria. The occasion was an entertainment honoring Mr. and Mrs. B. B. Foster. Said Parham:

[26] Conversation with Needham B. Weatherford, Camden, October 13, 1959; conversation of Nora Havard with Byron Pate, Camden, June 14, 1965. See also Inquest Book, Justice of the Peace, Camden, 1900–48.

[27] Conversation of Nora Havard with L. M. Phillips, Sr., Camden, June 21, 1965; W. T. Crouch, "The Negro in the South," in *Culture in the South*, pp. 434–35; conversation of Weldon Hunt with Pope Ware, Sarber, June 12, 1965.

[28] Conversations with Clyde J. Woodward, Sr., Nacogdoches, January 16, 1960, August 1, 1962; conversation with Needham B. Weatherford, Camden, October 13, 1959; conversation with Dr. George F. Middlebrook, Caro, October 15, 1959; conversation with George Morrison, August 6, 1958.

I can truthfully say you have a group of Negro laborers that will not join radicals for sinister motives of destroying the hand which is feeding them. They will give an honest day's work for an honest day's pay. They are always willing to give white people credit for the good things that they do for the race, and they do not blame all white people for the evils of some. We are very grateful to the Foster Lumber Company for giving us employment, that we may take care of our families and loved ones and not have to depend on Public Relief for a livelihood. We have shown loyalty to both our employers and our government, and a willingness to shed our blood that the Old Glory may forever wave o'er the land of the free and home of the brave. Never was there a Negro found guilty of espionage, sabotage, or other sedition. Being black, means above suspicion; being black means that you need no Dies Committee to investigate our unAmerican activities.[29]

Granting that Parham's remarks constitute rather fulsome praise—and that no doubt the local manager helped with the writing and the organization—the sentiments were representative of those of a large group of black lumber workers in Texas lumber towns in the generation following the turn of the century. Most of them, even the men in the bear pit, were loyal to the owners.

To most of the residents of the company town, especially the wives and children, the center of interest and activity was not the mill, with its lowering stacks and menacing whistle, but the commissary. The company store usually stood in the center of town, opposite the company offices. In a hundred company milltowns the commissary had a distinctive appearance: a large, two-story frame building with a false front that made it look even taller. In front was a long covered porch or veranda across the entire width of the building. Inside, the first floor was devoted to merchandise and an office in the rear. The second floor was used for storage. This description, of the company store at Camden, would fit equally well the company stores at Keltys, Diboll, Willard, Fostoria, and elsewhere.[30]

In comparison with the average general store in an East Texas village, the company commissary was a large, prosperous operation. Many company stores regularly carried $25,000 to $50,000 worth of stock and did an annual business of $100,000 to $150,000. Since the mill hands and their families usually bought all their needs from the company store, it carried virtually everything, and anything that it did not have in stock it would order. A well-stocked commissary carried everything from candy to coffins, from hats to horse collars, and from yams to yard goods.[31]

The interior of the company store also generally conformed to a pattern. Goods were usually (but not always) neatly piled in shelves or on display counters. Hat boxes were stacked high, and other items hung on the wall. Groceries were arranged in a separate section. When electricity arrived, the Texas heat was somewhat relieved by large rotating ceiling fans. Near the door was a long, uncluttered counter where the goods were laid out, checked, and paid for, and, if necessary, wrapped. In the rear, or sometimes on the balcony, was the manager's office, where accounts were kept and orders for merchandise were filled out for the wholesale houses. The frequent comment of visitors was one of surprise that such a large and varied stock could be found in such an isolated place.[32]

[29] Landy Parham, "Sentiments of the Negroes of Fostoria to the Executive Board of the Foster Lumber Company," December 3, 1940, Forest Lumber Company Papers, Forest History Collections, SFA Library.

[30] See pictures of commissaries in *American Lumberman*, January 18, 1908, and September 26, 1908.

[31] Commissary Records, 1904, Foster Lumber Company Papers; conversation with Needham B. Weatherford, Camden, October 13, 1959; conversation with Clyde J. Woodward, Sr., Nacogdoches, August 1, 1962; conversation with W. S. Brame, Livingston, August, 1962.

[32] Often the workers could not afford to buy coffins for their deceased, even though the coffins the store carried in stock were usually relatively inexpensive. Most workers and family members were buried in coffins made from company lumber in the cabinet shop. The women of the community padded and lined the coffin with materials purchased from the company store. The dead worker was laid out in cotton trousers and a shirt, almost never in a suit. Since almost no company towns had a mortician,

Company office. The office manager was often paymaster, rent collector, and banker. *Courtesy of* *Stephen F. Austin State University, Forest History Collections, Thompson photo albums.*

The commissary. *Courtesy of Stephen F. Austin State University, Forest History Collections,* *Thompson photo albums.*

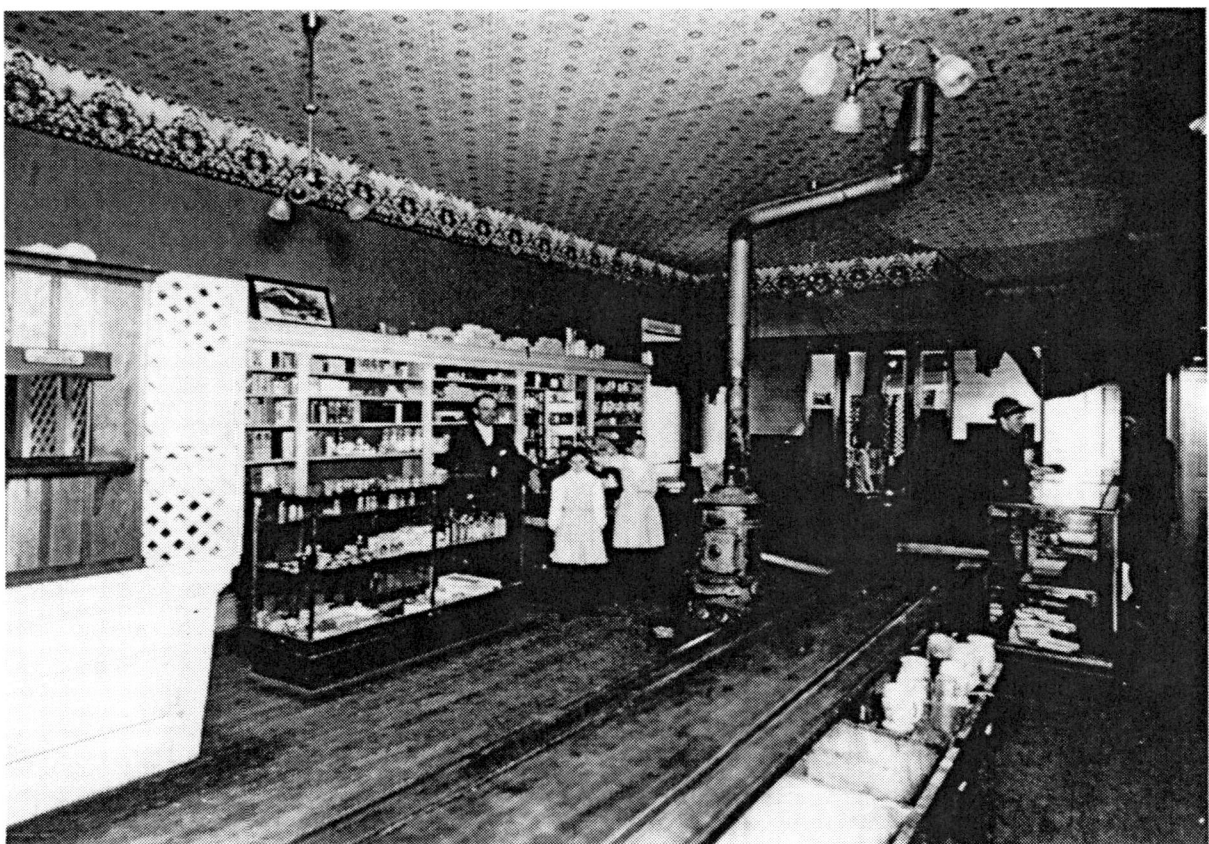

Interior of the commissary. *Courtesy of Stephen F. Austin State University, Forest History Collections,*

Thompson photo albums.

For the manager and his assistants the work was not particularly hard, but the hours were long. The commissary opened early, often at seven o'clock in the morning, and remained open until the last load of workers had returned from the woods at night—sometimes as late as eight or nine o'clock, or, if there had been trouble on the railroad, even later. It was not unusual for the commissary workers to go home to supper in rotation and return for another two or three hours' work.[33]

A break in the monotony doubtless came for the commissary manager with the arrival of a traveling salesman, or drummer, who represented a manufacturer or a large wholesale house in Houston, Galveston, Shreveport, or even Saint Louis. Not only did he bring samples and catalogues of his line of goods, but he brought news and gossip from the outside world. Since the drummer usually traveled up and down a given railroad line, calling on the milltowns along the route, he also brought news and gossip from neighboring mills. This enlivened the visit and kept one milltown fairly well informed of activities in other towns in the Piney Woods.[34]

---

the body was prepared for burial by friends of the deceased, often aided by the manager of the store, who over the years became a sort of jack-of-all-trades. Conversation of Nora Havard with Ida Baker, Camden, June 13, 1965; conversation with L. M. Phillips, Sr., Camden, June 21, 1965; conversation of John Larson with Mrs. Frank Farrington, Diboll, 1965.

[33] *American Lumberman*, September 26, 1908; conversation with W. S. Brame, Livingston, August, 1962; conversation with Needham B. Weatherford, Camden, October 13, 1959; conversation with Clyde J. Woodward, Sr., August, 1962; conversation of John Larson with Mrs. Frank Farrington, Diboll, June, 1954.

[34] The ACLCo. Papers contain many letters regarding orders given to visiting salesmen and asking for salesmen to call. Before 1900 salesmen from Shreveport called about as frequently as did those from Houston or Galveston. Then the Shreveport market dwindled owing to discriminatory freight rates until the settlement of the *Shreveport* case in 1914. See *Houston, East and West Texas Railway* v. *United States*, 234 U.S. 342 (1914).

Typical orders to wholesalers from the commissary provide clues to the life and customs of the time. Orders for galvanized washtubs, washboards, lye soap, and washbowls are evidence of the backbreaking toil of the women of the company town. Orders for cases of snuff, cuspidors, and "Star Navy" and other brands of tobacco point to the widespread habit in East Texas of chewing rather than smoking tobacco. Orders for castor oil, Epsom salts, diarrhea balsam, worm candy, and Grove's Chill Tonic make clear some of the common ailments of the people of the company town and their efforts to treat themselves. There are also more exotic orders, such as a case of oysters, boxes of oranges, sardines, sausages in oil, cigars, and expensive hats and shoes. These items were doubtless for the managers and for special occasions.[35]

The daily diet of the logger and lumber worker was simple and predictable. In a survey made in 1901 the United States Bureau of Labor Statistics sought to determine the eating habits of the workingman in the various regions of the United States, including the south-central region, composed of Louisiana, Arkansas, Oklahoma, and Texas, including, of course, East Texas lumber workers. In a comparison with workingmen in the five other regions, workers in the south-central region were fifth in consumption of fresh beef and sixth in consumption of salt or cured beef. On the other hand, they ranked first in consumption of salt pork and second in consumption of fresh-pork products. They ranked sixth in consumption of other meats, poultry, and fish. They ranked fifth in consumption of milk, sixth in consumption of butter, and about average in egg consumption. They consumed more lard and far more molasses than did workmen in other sections, and they drank less tea but more

coffee than workers in any other area. They ranked first in the consumption of rice and about average in consumption of potatoes.[36]

The survey indicated that the East Texas lumber worker subsisted largely on a diet of salt pork, starches, corn bread, foods fried in large amounts of lard, sweetened with molasses and washed down with quantities of cheap coffee—Arbuckle's or Rio. He added to this diet seasonal vegetables such as cabbage, beans, onions, and fresh corn raised by the women in the family garden plot. According to the Department of Labor, the south-central worker spent less for green vegetables than did workers in other sections and also purchased less fruit.[37] His diet was simple, hardy, filling, but lacking in both variety and needed vitamins and minerals (of whose existence he was blissfully ignorant). It was augmented some by seasonal garden vegetables but not as many as it should or might have been. Most of his food—salt pork, cornmeal, rice, potatoes, lard, molasses, and coffee—was purchased—at the company store.

The key to the operations of the company commissary was the merchandise check. Although the store frequently did an annual business in excess of $100,000, very little of it was for cash. Nor did the company store, except during layoffs and breakdowns, do much conventional credit business. One company estimated that it paid its employees more than 90 percent of their wages in merchandise checks and that the checks eventually returned to the company through purchases at the company store, rent, and other purchases. As could be expected, the merchandise check drove United States legal tender out of the company towns.[38]

The merchandise check was not a check at all in the conventional sense but a pseudo coin.

[35] ACLCo. to Wiener Loeb Grocery Co., Shreveport, March 7, 1901; J. H. Kurth to L. & H. Blum, Galveston, July 28, 1888; ACLCO. to P. J. Younger & Co., Homer, La., December 9, 1904; Commissary Inventory, 1902 (box 38), ACLCo. Papers; W. T. Carter & Bro. Commissary Receipt File, Camden, Texas.

[36] For prices, 1890–1912, see U.S. Department of Labor, Bureau of Labor Statistics *Bulletin* no. 132, p. 31, quoted in Allen, *East Texas Lumber Workers*, p. 121.

[37] Ibid., pp. 122–23.

[38] Speek, "Notes on Investigations of Three Texas Lumber Towns."

A collection of merchandise checks. *Authors' collections.*

It was usually made of a base metal—brass, aluminum, or tin alloy—with the name of the company stamped on one side and the amount on the other. The size of the check often varied with the denomination. Many companies used a hard-cardboard merchandise check with distinctive colors for the several denominations (red for 50 cents, blue for 25 cents, yellow for 10 cents, and green for 5 cents).[39] There were variations from this practice, however; some companies issued punch-out cards, coupon books, or merchandise books. Whether these substitutions for cash were in the form of brass coins or coupon books, they had the same essential characteristic: they were good only for merchandise at the company store or for other payments to the company.[40]

Technically, the company issued merchandise checks at the request of the worker, but since cash paydays were infrequent and irregular, the worker had to draw his pay in merchandise checks to live. Kirby for a time paid twice a year: on Christmas Eve and the Fourth of July. The Alexander Gilmer Lumber Company issued "pay checks" or pay slips every

evening at the end of work but had a payday only once a month, at which time the company redeemed pay slips for purchases at face value; independent merchants discounted them at 5 to 10 percent. The companies normally did not redeem merchandise checks for cash.[41]

The lumber worker and his wife regularly found that the prices at the commissary were a little higher than those at independent stores. A survey of comparative prices by the United States Department of Labor in 1914 under the direction of Peter A. Speek found that at Kirbyville the commissary prices were consistently about 10 percent higher than those at private stores. For example, 10 pounds of lard cost $1.65 at the commissary but $1.50 at independent stores. A thirty-five pound bag of cornmeal was $1.00 and $.90. Cottonseed meal was $1.75 and $1.60. The prices of other items varied in similar relation.[42] In Kirbyville the independent merchants regularly discounted the merchandise checks at 10 to 20 percent, since the workers could use them only in trading at the store, for the company would not redeem them in cash. Thus the worker and his wife found that his wages were regularly subject to a discount of at least 10 percent in purchasing power no matter where they purchased their necessities. The system, by its very nature, kept the worker broke.

In 1916 the commissioner of labor and industrial statistics of Louisiana described the company store in terms that were equally applicable to those of Texas and the other Gulf states. The price gouging was a perplexing problem, he conceded, and he recommended passage of legislation to regulate the operation of company-owned commissaries. He wrote:

On account of many industries, particularly among the lumber operators and sugar manufacturers, being located at outlying or isolated points where there are no other stores, it might become necessary to operate . . . a commissary, but at the same time

[39] In the Forest History Collections, Stephen F. Austin Library, is a collection of the most popular types of merchandise checks and also some that were relatively rare.

[40] Henderson and Kurth to Clark and Courts, Galveston, July 30, 1889; ACLCo. to George D. Benard & Co., St. Louis, May 4, 1894, ACLCo. Papers; conversation with Clyde J. Woodward, Sr., Nacogdoches, October 3, 1957; conversation with Needham B. Weatherford, Camden, October 13, 1959.

[41] Speek, "Notes on Investigations of Three Texas Lumber Towns."

[42] Ibid.

it does not follow such operations should be necessarily an evil. If such places were properly conducted, that is, sell laborers such goods and wares as were necessary and furnish same . . . at a small profit, no adverse comment would come from this office, but ordinarily these conditions do not exist. . . . The average company store . . . is operated as a money-making proposition and prices are fabulously high; . . . last July, while flour was selling throughout the state at about $11.50 per barrel, some commissaries were charging their employees as much as $19.00 for same, and at the same time were charging laborers 42¢ per pound for salt meat and 10¢ per pound for second grade rice, while the public market prices were about 28 and 6¢ respectively for these commodities.[43]

If the worker needed cash for any purpose, such as an emergency trip to a sick or dying relative, he had only two courses open to him. He could sell his accumulated merchandise checks to a private individual at a large discount—almost every company town had one or more loan sharks. Speek reported one such transaction in which a worker exchanged $2.60 in merchandise checks for $1.80 in legal tender. At this rate the worker suffered a 30 percent loss in wages. When the investigator asked whether such transactions were common, a company official replied, "Oh, quite often. The people are ignorant."[44] The second alternative for the worker in need of emergency cash was to draw a "time certificate" (if he had any accumulated time) from the company office. The certificate could be cashed at the local bank at a 10 to 20 percent discount, regardless of the maturity date.[45] Thus except on payday the worker had to submit to excessive interest or discount charges if he needed United States currency. The commissary and merchandise checks effectively kept all of the worker's and his family's transactions within the company system.

Many company towns had detached "suburbs" at the logging front, some as far as twenty miles from the mill and the main establishment. Most logging camps were merely temporary settlements where the men lived in converted railroad boxcars for a few weeks or built shacks and other temporary structures. Some logging camps, however, housed several hundred persons, including wives and children, and took on the appearance of permanence. One such semipermanent logging town was Camp Ruby, in southern Polk County, a W. T. Carter & Brother Lumber Company property connected with Camden by a mainline tramroad. Another was Fastrill, in western Cherokee County, built by the Southern Pine Lumber Company in 1922 and connected with Diboll by the Texas South-Eastern Railroad. A third was Camp Worth, in San Augustine County, owned by the Frost-Johnson Lumber Company of Nacogdoches and constructed soon after World War I on the main line of the company's railroad, the Nacogdoches and Southeastern. These towns were built when logging operations extended so far from the mill and town that it was cheaper and more efficient to build a logging town than to transport the workers back to town every night. Each of these towns, and no doubt a number of others, had an existence of ten to twenty years and enjoyed facilities similar to those in the parent company town. For example, Fastrill had a school, two churches, a post office, and a branch commissary. It claimed a peak population of about six hundred. As the timber was cut out, these logging towns "at the front" were abandoned, and the families moved back to the principal company settlement.[46]

Despite the inequities of the merchandise-

[43] Louisiana Department of Commissioner of Labor and Industrial Statistics, *Ninth Biennial Report* (Louisiana, 1915–16); quoted in Allen, *East Texas Lumber Workers*, pp. 160–62.

[44] Speek, "Notes on Investigations of Three Texas Lumber Towns."

[45] Ibid.; Creel, "Feudal Towns of Texas"; conversation with J. C. Smith, Wiergate, July 30, 1963.

[46] *Nacogdoches Daily Sentinel*, September 16, 1965; conversation with Clyde J. Woodward, Sr., Nacogdoches, October 3, 1957; conversations with Needham B. Weatherford, Camden, July 25, 1958, October 13, 1959; conversation with Clyde Thompson, Diboll, May 10, 1965. The name *Fastrill* was coined from the names of three Diboll officials, F. F. Farrington, P. H. Straus, and Will Hill. Bob Bowman, "Touring East Texas," *Nacogdoches Daily Sentinel*, September 16, 1965.

The company store at Fostoria. *Courtesy of Stephen F. Austin State University, Forest History Collections.*

check system, the excessive prices charged by the company store to its captive customers, the infrequent paydays, and the poor housing, the sawmill workers were surprisingly loyal to their employers. In the entire history of Texas lumbering no band of workers burned a mill, picketed a main office, or sacked a company store. Instead, most of the veteran employees—even those who had held only manual-labor jobs—usually had high praise for the millowners. This spoke well for the personal relationships that such men as Thomas L. L. Temple, Joseph Kurth, W. T. Carter, E. A. Frost, and John Henry Kirby had with their employees and their families.

In some respects there was much to be said for these entrepreneurs who established and operated the company towns in East Texas. They created an industry, provided jobs, built homes, and organized a stable society where none had existed. In comparison with the rural villages and the red-clay and sandy farms in the Piney Woods, the standards of wages, hours, housing, and opportunities for advancement were more favorable in the company town. Even the pernicious merchandise-check system probably was begun without malice but as a measure to help the company continue to operate without spending large amounts of cash. Actually, of course, the merchandise-check system was not necessary for the solvency of the company. After 1900, Lutcher and Moore, one of the oldest and most successful of the Texas lumber companies, paid its men in cash every Saturday night.

The company-town system has been denounced as vicious and pernicious, enslaving and degrading to the workers and their families. It was all these and became more so as years and decades passed without any change in the social, economic, or political control of the community. Too often the company not only exploited the worker and his family during the years of operation but then suddenly left them without a future with the ending of logging and the closing of the mill. As one veteran lumber employee remarked, "The saddest sound in the world is a silent mill whistle."[47]

Yet many lumber owners were genuinely concerned about the well-being of their employees and consistently worked to improve working and living conditions. Three generations of Carters, two generations of Kurths and Hendersons, and three generations of Temples disproved the stereotype that sawmill employer-employee relations were all exploitation and hostility. These companies were still in operation after more than seventy-five years of lumbering. Their company towns

were steadily improved, and their labor turnover was at a minimum.

The Texas lumber-company town was a characteristic of the generation before World War I. As the automobile, improved roads, and advanced means of communication entered the Piney Woods, the isolation and dependence of the worker disappeared. Company-town organization lingered, to be sure, for another two decades, but it lost its feudal and paternal characteristics. The destruction of the virgin forests, the coming of the Great Depression, and the desertion of many workers for the East Texas oil fields hastened the abandonment of many company towns and milling operations. In those that remained, the companies discontinued the use of merchandise checks (especially after 1937), encouraged their employees to buy their own homes, and made prices at the company store competitive with those in the county seat. Doubtless the company-town, company-store pattern of economic and social organization filled a real need when the commercial lumbering industry first moved into Texas, but these institutions lingered long after the need had passed.

---

[47] See Allen, *East Texas Lumber Workers*, p. 199, for a critical appraisal of the system. Conversation with Dr. George F. Middlebrook, Caro, October 15, 1959.

CHAPTER 10

The Bonanza Era

There is a tide in the affairs of men which
taken at the flood, leads on to fortune.—
William Shakespeare

IN the generation following the Civil War, when the businessmen of the East and the Great Lakes states thought of Texas—if they thought of Texas at all—they imagined wide-open spaces, cattle country, and arid grasslands. The great forested region of East Texas, as large as many states, remained largely ignored by most, despite articles and travel accounts that visitors to the region regularly published in the eastern press. Upon investigation, as has been discussed earlier, the great stands of pine impressed lumbermen from "older" sections of the country. It was truly a bonanza opportunity that beckoned, and as in the gold discovery in California and the oil strike at Spindletop, entrepreneurs from throughout the country hastened to exploit it.[1]

This period, which lasted from 1876 to 1917 and has been called "Victorian Texas," was also a profitable era for entrepreneurs in many other fields. The nation turned from the problems and animosities of the Civil War and Reconstruction to a new era of expansion and growth. These were the years of rapid construction of railroads, and by 1901 the rail net was virtually complete. Railroad mileage in Texas reached ten thousand miles in that year. The completion of the telegraph system at the same time provided business and industrial

leaders with quick, cheap, reliable communication to all parts of the state. It is not surprising that the greatest growth of the lumber industry came in those years.[2]

They were also good years for businessmen, politically and constitutionally. Not only were most presidents, senators, and congressmen strongly business-oriented, but federal judges stressed property rights, freedom of contract, and individual enterprise. In Texas most governors and legislators shared the same economic philosophy.[3] Until the advent of Theodore Roosevelt in 1901 such laws as existed regulating interstate commerce and enforcing anti-trust policy were largely ignored. There was no income tax, and owners augmented their profits as they could. In short, the general outlook of the period encouraged entrepreneurs to flourish according to the tenets of the classical economics of the nineteenth century.[4]

Most of the Texas lumber barons probably had never read Herbert Spencer or William G. Sumner, but they were, nevertheless, staunch

[1] Frank H. Taylor, "Through Texas," *Harper's New Monthly Magazine* 59 (October, 1879): 706; James Boyd, "Fifty Years in the Southern Pine Industry," *Southern Lumberman* 144 (December 15, 1931): 59–67.

[2] John S. Spratt, *The Road to Spindletop: Economic Change in Texas, 1875–1901*, pp. 32–33, 250–66; Charles P. Zlatkovich, *Texas Railroads: A Record of Construction and Abandonment*, pp. 5, 37.

[3] Thomas C. Cochran and William Miller, *The Age of Enterprise: A Social History of Industrial America*, 2d ed., pp. 119–50; Rupert N. Richardson, Ernest Wallace, and Adrian Anderson, *Texas: The Lone Star State*, 4th ed., 268–78, 346–54.

[4] Cochran and Miller, *The Age of Enterprise*, pp. 152–53; Harold U. Faulkner, *Politics, Reform, and Expansion, 1890–1900*, pp. 13, 101–102.

W. T. Carter. *Courtesy of Stephen F. Austin State University, Forest History Collections.*

individualists and believers in laissez-faire. There is no record that the people of any of the East Texas counties ever considered forming a cooperative organization to manufacture and market lumber to the profit of the general community.[5] Instead, even the native sons who developed large mills followed the example of entrepreneurs who came to the Piney Woods from other sections. In general they were aggressive, determined, stubborn men, dedicated to turning the Texas pines into merchantable lumber and that lumber into dollars. Their practices did not differ markedly from those of tycoons in oil, steel, or railroading. They

treated their mills, their vast acreages, their company towns, and their employees as though they were feudal baronies, which perhaps they were.[6]

The lumber barons were not without their problems, however. They faced a very unstable market in which prices fluctuated widely and often broke without warning. For example, the price for average mill-run lumber varied from $12 a thousand board feet to more than $18 a thousand between 1906 and 1913. Between 1924 and 1933 the fluctuation in the same-quality lumber ranged from a high of $30 to $14 a thousand.[7] With production costs relatively stable, such variations could prove disastrous to many millowners. In addition the southern-pine lumber manufacturer had to work against the prejudices of northern and western builders, who were used to working with white pine. Only reluctantly did they accept yellow-pine lumber as equal in strength and versatility.

The lumber manufacturer was plagued with a constant shortage of freight cars, slow deliveries, and unfavorable freight rates in comparison with those of the Great Lakes states and the Pacific Coast. Most of the railroads that penetrated the Piney Woods were small lines that found it difficult if not impossible to provide cars as demanded by the mills. If business was brisk, the lumberman could not get cars to ship his orders as fast as he could fill them; if business was slow, he found that the railroad was demanding that he move the cars on the siding or release them. Company bookkeepers and secretaries spent much of their day corresponding with railroads over cars, misdirected shipments, demurrage, division of rates, and shortages. It is not surprising that many lumbermen sought alternate rail outlets for their

[5] There was one lumber company in Rosenberg, Texas, which called itself the "Independent Co-operative Lumber Company," but it was only another commercial venture engaged in the wholesale and retail lumber trade. See ACLCo. Papers, Forest History Collections, Stephen F. Austin State University Library, Nacogdoches; hereafter cited as SFA Library.

[6] Ruth A. Allen, *East Texas Lumber Workers: An Economic and Social Picture, 1870–1950*, pp. 19–50.

[7] "Hearings by the Federal Trade Commission on Condition of the Lumber Industry in the South," *American Lumberman*, July 24, 1915; Peavy-Moore Lumber Company, Statement of Operating Costs, 1924–39, in ACLCo. Papers.

products in hopes of improving their bargaining position.[8]

Weather was also a factor in the Piney Woods. In the winter and spring the ground was frequently too wet for the loggers to work for weeks at a time. In the summer the blazing Texas sun was enervating to both men and beasts. Then there was the constant danger of fires. Not only were there mill, kiln, yard, planer, boiler, and warehouse fires, but the woods were often afire from sparks of wood-burning locomotives, lightning, accident, or deliberate arson. Petulant "nesters" often set incendiary fires out of resentment toward big companies that had deprived them of their favorite hunting areas. Large, impersonal companies with absentee owners were harder hit by incendiaries than were smaller, local firms.[9]

Not all companies made money. Such an example was the Central Coal and Coke Company of Kansas City, Missouri. The "4-C," as it was called locally, came to Texas soon after the turn of the century and in 1902 built a large plant near Ratcliff, in Houston County. The mill was a most ambitious project. It was described as a triple-band rig (with one extralong carriage) and a fifty-two-set gang saw (a very large saw with fifty-two straight up-and-down saws). It was rated at a capacity of 300,000 board feet an eleven-hour day. In addition to the mill proper, the company built a planing mill, dry kilns, sheds, shops, and a short-line railroad, the Eastern Texas, which connected with the Cotton Belt. The parent company, presided over by Charles S. Keith, had operations in Arkansas, Oklahoma, and Louisiana, as well as in Missouri and Texas.[10]

The company had difficulty purchasing sufficient timberlands to support this large-scale operation. Situated in a triangle surrounded by Kurth, Temple, and Thompson properties, the 4-C bought lands and stumpage wherever they could be found, even in scattered lots. The older companies competed for available timber, and on at least one occasion a 4-C timberman turned out to be a double agent who was also buying desirable timber for one of the established companies. Although the 4-C managed to acquire about 120,000 acres, by 1918 the company had cut out. The company also pursued an arbitrary policy with its employees. In an effort to force them to trade entirely at the company commissary, the manager had a high fence erected separating the company properties from the older town of Ratcliff. Much hostility was generated against the company, which suffered many annoyances, spite fires, bombings, and petty reprisals.[11]

In 1918 the 4-C abandoned operations and sold its properties to the Houston County Timber Company. The cutover land would not support such an elaborate lumber-manufacturing plant, so the new owners dismantled the big mill, replacing it with a more modest outfit. Soon all that was left of the ambitious 4-C operation were the sturdy reinforced-brick foundations, the millpond, and the severely cutover land. The 4-C had suffered from late entry into lumber manufacturing along the upper Neches. It faced stiff competition from well-established rivals, and it was never able to acquire an adequate timber base for sustained large-scale operations. In addition it suffered the disadvantages of being a foreign (out-of-state) corporation with absentee direction and arbitrary management

---

[8] See S. W. Henderson to A. S. Dodge, August 4, 1894; S. W. Henderson to R. H. Bowron, December 7, 1900; ACLCo. to R. C. Hancock, September 27, 1900; ACLCo. to Burton Lingo Co., October 1, 1900; Henderson and Kurth to A. C. Petri & Bros., March 13, 1890; Henderson and Kurth to H. D. Milton, July 18, 1888, ACLCo. Papers.

[9] For an account of local hostility toward large corporations see Ed Kerr, "Southerners Who Set the Woods on Fire," *Harper's* 217 (July, 1958):28–33.

[10] *American Lumberman*, November 1, 1902; conversation with H. Z. Collier, Kennard, Texas, July 29, 1964, Oral History

Collections, Forest History Collections, SFA Library. Unless otherwise noted, interviews and conversations cited are in these collections, and all interviews were conducted by principal author Maxwell. Apparently the corporate name for the Texas operation was the Louisiana and Texas Lumber Company, but it was chiefly known as the 4-C. *American Lumberman*, April 4, 1908, p. 66.

[11] Conversations with H. Z. Collier, with Jim McKinney and with Gary Mahoney, Ratcliff, and Kennard, July 29, 1964.

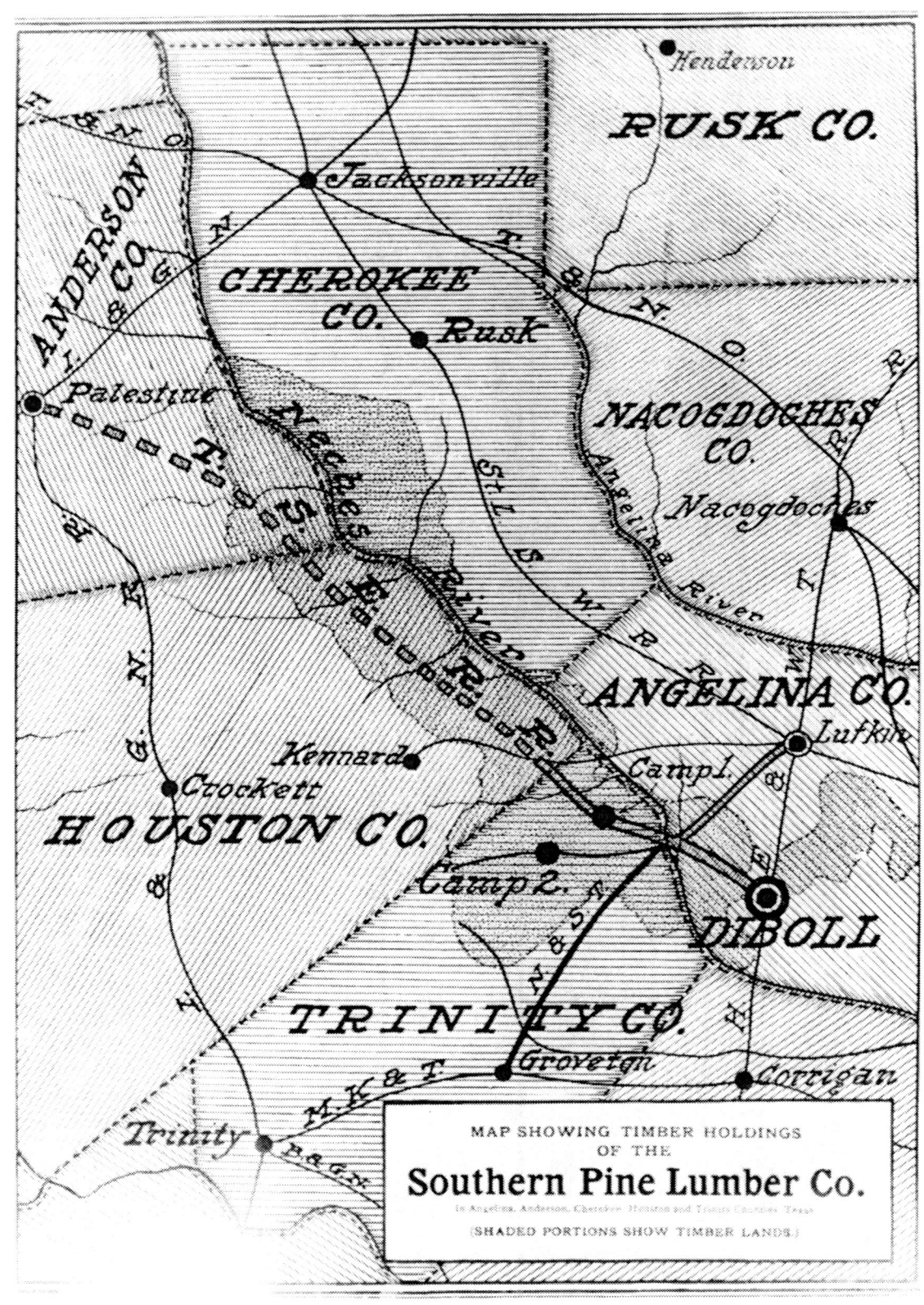

Southern Pine Lumber Company operations. *Authors' collections.*

policies. It is small wonder that the 4-C mill was short-lived.[12]

The total number of lumber plants in East Texas at any given time would have been difficult if not impossible to determine. They appeared and disappeared according to the economic climate like mushrooms after a rain. The United States census of 1900 reported 637 lumber-manufacturing establishments in Texas with an invested capital of $19 million, paying $3,094,127 in wages to 7,924 employees. Of these about 200 were permanent mills capable of cutting as much as 50,000 board feet a day, and fewer than half of these mills could be described as big mills with complete lumber-manufacturing facilities and a capacity of 100,000 board feet a day. The rest were smaller mills ranging down to the two-man portable mills that farmers might bring out in times of good prices and slack farm work. These so-called woodpecker or peckerwood mills usually manufactured only rough lumber of an inferior grade that was sold cut-rate and consequently depressed the market. Often the small mills sold their products directly to larger mills for remanufacture and sale. In slack times the small mills disappeared when their owners returned to farming, ranching, or other occupations.[13]

In 1910 spokesmen for the industry estimated that there were about 625 lumber mills in the state, "about 375 to 400" of which were of a size of any importance. Basing their figures on the 1920 census, observers reported about 450 lumber mills in East Texas but estimated between 400 and 500 unreported woodpecker mills in the timber belt. The *Texas Almanac, 1931,* reported about 160 lumber mills in the state, not counting small, irregularly operated portable plants.[14]

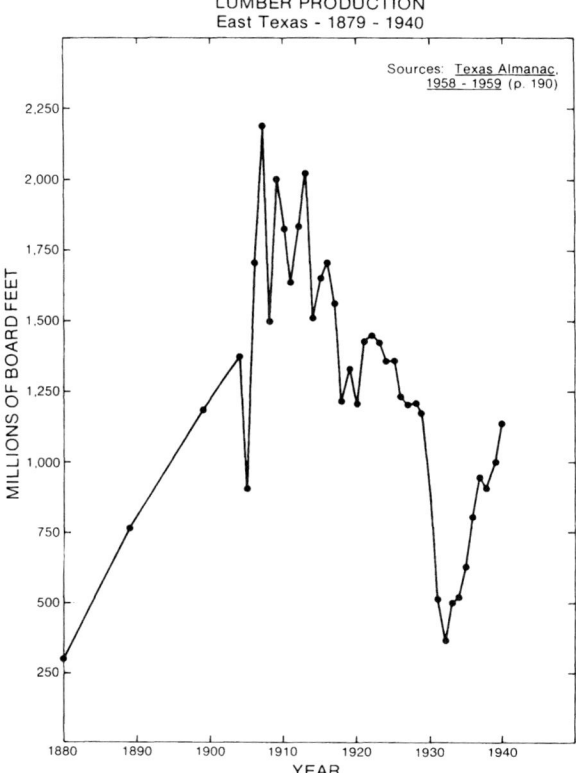

Lumber production in Texas, 1879–1940. *Courtesy of the Department of Forest Science, Texas A&M University.*

Even these figures, however, are misleading. Most of the lumber and virtually all of the quality lumber was manufactured at the larger, permanent mills. In 1907, the year in which Texas achieved its all-time record cut of almost 2.25 billion board feet, the *American Lumberman* itemized the cut of 99 reporting mills (apparently the leaders in the industry) at 1,257,600,000 board feet. That would leave less than 1 billion feet cut by the remaining 400 to 500 mills. And this was a boom year, which brought out the "sometime peckerwoods."[15]

The irregular annual lumber production in Texas, shown in graph form, approximates a

[12] The 4-C timberlands were eventually included in the Davey Crockett National Forest; see U.S. Forest Service leaflet on Davey Crockett National Forest, Forest History Collections, SFA Library. See also conversation with Simon W. Henderson, Jr., August 10, 1964, Keltys.

[13] *Texas Almanac,* 1904, pp. 143–44; *American Lumberman,* May 17, 1902; conversation with Ernest L. Kurth, 1959.

[14] *Texas Almanac,* 1910, p. 111; *Texas Almanac,* 1926, p. 179; *Texas Almanac,* 1981, p. 126.

[15] *American Lumberman,* April 4, 1908, p. 66. Some individual production figures for the year 1907 may be interesting: Lutcher and Moore reported 65 million board feet; 4-C also cut 65 million board feet, Southern Pine Lumber Company, of Diboll, cut 53 million board feet, and Angelina County Lumber Company cut 16,834,000 board feet, operating only seven months. Thirteen mills of the Kirby Lumber Company reported a total of 318,943,000 board feet.

General view of sawmill site, Frost Mill, Nacog-
doches. *Courtesy of Stephen F. Austin State Univer-* *sity, Forest History Collections.*

ragged pyramid. In the fifty years of the bonanza period, from 1880 to 1930, the annual
cut of pine lumber rose from a modest 327 million board feet in 1880 to more than 1 billion
board feet in 1900 and to an all-time peak of
2,197,233,000 board feet in 1907 (and an estimated third-place rank among the nation's
lumber producers in that year). Three times between 1907 and 1916 the Texas manufacturers
produced more than 2 billion board feet a year.
That period could be considered the "golden
decade" of Texas lumbering, for during this
time the annual cut never fell below 1.5 billion
board feet and averaged more than 1.75 billion
board feet for the entire decade. Never again to
date have Texas lumbermen produced as many
as 2 billion board feet a year, and the annual
cut declined slowly through the 1920s until
1930, when the total fell below 1 billion board
feet.[16]

The lumbermen soon learned that "in union

there is strength." They organized trade associations, which were useful in establishing lines
of communication, exchanging ideas, forming
rallying points for concerted stands on public
policy, and even on occasion controlling the
market. The association memberships ranged
from national to regional and state and often
represented highly specialized interests within
the industry. Among the better-known organizations were the Southern Pine Association,
the Lumbermen's Association of Texas, the
Southwestern Lumbermen's Association, the
Yellow Pine Manufacturers' Association,
the Southern Lumber Operators' Association,
the Southern Cypress Manufacturers' Association, the National Association of Box Manufacturers, the National Hardwood Lumber
Association, the National Association of Lumber and Sash and Door Salesmen, the Lumber
Secretaries' Bureau of Information, the National Wholesale Lumber Dealers' Association, the Forest Products Association, and the
American Forestry Association. In 1916 the

[16] *Texas Almanac, 1958–1959,* p. 190.

American Lumberman (the national publication for the industry) listed forty-three national, state, and local meetings scheduled for that year.[17]

The most important trade association for Texas and Louisiana was no doubt the Southern Pine Association. This energetic and prestigious organization was preceded by several short-lived regional agencies that were relatively ineffective and frequently under fire for violations of antitrust laws. In 1883 lumbermen along the Arkansas, White, Black, and Saint Francis rivers formed the Missouri and Arkansas Lumber Association for the purpose of standardizing their sizes and upgrading their products. Texas and Louisiana lumbermen refused to affiliate with this organization until 1899, when a compromise was reached on a uniform gauge of sizes and grades. In 1906 the name of the group was changed to the Yellow Pine Manufacturers' Association (YPMA) and in another year or two its membership totaled 300 companies, including most of the larger manufacturers in the central and Gulf South and representing about one-third of the total annual Southern-pine production.[18]

The YPMA was soon under fire for violating both federal and state antitrust laws. In 1907 Texas Attorney General R. V. Davidson demanded a legislative investigation of the lumber manufacturers, who, he charged, were fixing prices through their regional and national organizations. A resolution was introduced in the United States Senate to authorize the secretary of commerce and labor to investigate the lumber manufacturers and their trade associations to determine whether a trust existed that restrained lawful trade and controlled the price of lumber. Finally, in 1914, the Missouri Supreme Court found a number of the members of the YPMA guilty of limiting the output of yellow-pine lumber and fixing prices contrary to the state's antitrust laws. Shortly afterward the member companies withdrew from

the organization, and the Yellow Pine Manufacturers' Association became defunct.[19]

During the same years the Southern Lumber Operators' Association (SLO) was also active. This organization was composed of many of the same individuals and companies that held membership in the YPMA, and individuals and the press frequently confused the two. The SLO, however, included only larger manufacturers and apparently devoted most of its attention to identifying union organizers and countering efforts by unions to gain a foothold in the South. In this the SLO was largely successful. Not only did it defeat unionization attempts of the Brotherhood of Timber Workers to organize the workers in Louisiana and Texas mills and logging camps, but it maintained close contact with the member companies by means of bulletins and newsletters.[20]

Adopting the slogan "No Compromise—No Recognition," the SLO was effective in dealing with union organizers and any liberal-minded owners who wished to recognize the Brotherhood of Timber Workers.[21] By threats of boycott and expulsion from the organization they kept the lumber manufacturers in line, and they made good their threat to blacklist all union workers. The association provided the member companies with blank forms to be filled out by job applicants, authorizing their former employers to furnish their complete records to the new company, including any unfavorable or derogatory material. Any mention that the worker was a "Timber Worker" or even sympathetic to the union was usually sufficient to have him blacklisted.[22]

[17] *American Lumberman*, April 7, 1906, p. 41; *American Lumberman*, January 16, 1916, p. 44.
[18] John M. Collier, *The First Fifty Years of the Southern Pine Association*, p. 36.

[19] Ibid., p. 41; Ralph C. Bryant, *Lumber: Its Manufacturing and Distribution*, pp. 329–33; James Boyd, "Fifty Years in the Southern Pine Industry," *Southern Lumberman* 144 (December 15, 1931): 59–67.
[20] Southern Lumber Operators' Association to Angelina County Lumber Company, April 15, May 12, 1912, ACLCo. Papers; Vernon H. Jensen, *Lumber and Labor*, p. 86; Allen, *East Texas Lumber Workers*, pp. 165–90.
[21] George T. Morgan, Jr., "No Compromise—No Recognition: John Henry Kirby, the Southern Lumber Operators' Association, and Unionism in the Piney Woods, 1906–1916," *Labor History* 10 (Spring, 1969): 165–90.
[22] Grogan Manufacturing Company to Angelina County

From 1908 to 1912, the years of greatest union activity in the Southwest, the SLO maintained a system of "committees of correspondence," which quickly alerted individual members at the first indication of union activity in the area. As mentioned earlier, the SLO won a complete victory over the Brotherhood of Timber Workers and its parent organization (from 1912 on), the Industrial Workers of the World. No mill in the Texas-Louisiana area was unionized, and after 1914 the brotherhood declined and died. With the demise of this threat, the SLO also became inactive.[23]

The death of the Yellow Pine Manufacturers' Association had left southern lumber manufacturers without a central organization. To fill this void, a number of leading lumbermen met in New Orleans determined to form a stronger, more effective organization that would not be in conflict with federal or state antitrust laws. The result was the Southern Pine Association (SPA). It differed from previous associations in that it was incorporated under the laws of Missouri (and was thus in a sense placed under the watchful eye of a hostile court), members paid a subscription fee based on their annual production, and the association provided services and statistical information as the members demanded.[24]

From the beginning the larger manufacturers strongly supported the SPA. Charles S. Keith, of the Central Coal and Coke Company of Kansas City, became the first president, and John Henry Kirby was one of the first vice-presidents. Among the directors were W. H. Sullivan, of the Great Southern Lumber

E. A. Frost, principal owner of the Frost-Johnson Lumber Company, with mills in Arkansas, Louisiana, and Texas. *Courtesy of Stephen F. Austin State University, Forest History Collections.*

Company of Bogalusa, Louisiana; C. D. Johnson, of the Frost-Johnson Lumber Company, of Saint Louis; and T. L. L. Temple, of the Southern Pine Lumber Company, of Texarkana and Diboll. Among the active members and committeemen were Henry B. Hardtner, of Louisiana (known as the father of forestry in Louisiana and promoter of the "tree-farm" project), Robert A. Long, president of Long-Bell, and W. A. Pickering, of the Pickering Lumber Company, of Kansas City. Within a year the membership of the SPA represented an annual production of more than 6 billion board feet of lumber, which was more than half the total yellow pine produced in the Gulf South.

From the first year of its organization the SPA engaged in extensive trade promotion, seeking to improve the image of southern yellow-pine lumber in the eastern and northern markets. It provided a dealer's handbook and a book on farm plans and sponsored

Lumber Company, June 6, 1907. Occasionally, when the owner was sure that the worker "had learned his lesson" and promised to have nothing more to do with the union, he would be taken back. See Ernest Kurth to J. H. Kurth, Jr., September 7, 1911.

[23] SLO to ACLCo., April 15, 1912; SLO to ACLCo., May 22, 1912; Eli Weiner to C. P. Meyer, Kirby Lumber Co., July 7, 1911, ACLCo. Papers.

[24] John M. Collier, *The First Fifty Years of the Southern Pine Association*, pp. 41–43. For a comprehensive account of the Southern Pine Association see James E. Fickle, *The New South and the "New Competition": Trade Assocation Development in the Southern Pine Industry.*

T. L. L. Temple. *Courtesy of Stephen F. Austin State University, Forest History Collections.*

schools for salesmen. In addition, the SPA served as a clearinghouse for information and provided a great variety of statistical data for its members. Regularly the SPA undertook to speak for the industry to government, customers, and the general public.

During World War I the SPA coordinated the industry's war efforts. Through the Southern Pine Emergency Bureau, the War Service Committee, and the War Committee the industry produced and shipped several billion board feet of lumber and ships' timbers for government use (see chapter 12). In August, 1918, the SPA summarized the industry's contribution in this fashion:

Nearly all Army cantonments, camps and shore-houses in this country are of Southern Pine. Hundreds of ships are being built of Southern Pine exclusively. The government specifies it for boats and barges. Our battleships had decks of Southern Pine. Docks, piers and bridges are supported by Southern Pine timbers. Army supplies sent abroad are stored in Southern Pine buildings and our boys in

training in France are being housed in Southern Pine Shelters.[25]

In 1922 the SPA embarked on a crusade for standard "grade-marked" lumber. Earlier associations had devised and announced standards, but they varied greatly from state to state, and a great number of mills refused to be bound by them. Rules for export lumber were largely local. Perhaps because of the government's experience in the war, the United States Department of Commerce, under Herbert Hoover, took the lead in bringing about uniform standards for lumber. After many committee hearings, the department adopted the "American Lumber Standards" and urged all associations to adopt them or some variation of them. The Southern Pine Association adopted the American Lumber Standards and set up machinery to ensure that lumber sold on the market would be grade-marked, species-marked, and mill-indicated. By 1924 the SPA was ready to put the program into operation, and on July 1 of that year all members produced for general sale only lumber grade-marked as prescribed. The legend ⑩ SPA B&B on the end of a lumber product became the guarantee of its quality and inspection.[26] In 1940 the Southern Pine Inspection Bureau was established from the SPA to take over the long-term program of providing uniform grade standards for southern-pine lumber.[27]

Like its predecessor, the Southern Pine Association faced continuing investigations by the United States Department of Justice and the Federal Trade Commission for alleged viola-

[25] Collier, *The First Fifty Years of the Southern Pine Association*, pp. 63–68.

[26] The circled number 10 indicates the mill, in this case Lutcher and Moore; SPA shows that it has been inspected by the Southern Pine Association; and B&B means "B and Better," a top "select" classification for general yard lumber. Other grades in the select classification include C, C&Btr. and D. Common lumber is numbered 1, 2, 3, and 4. Southern Pine Inspection Bureau, *Short Course in Grading*; Boyd, "Fifty Years in the Southern Pine Industry," pp. 59–67; Collier, *The First Fifty Years of the Southern Pine Association*, pp. 80–84; *Gulf Coast Lumberman*, December 1, 1924.

[27] Southern Pine Inspection Bureau, *Short Course in Grading*, p. 1.

tions of the antitrust laws. During the Wood-row Wilson administration the FTC investi-gated the activities of the association and announced that evidence of price fixing, price regulation, and production restriction had been uncovered and turned over to the Justice Department. In 1922 suit was brought against the association charging combination, re-straint of trade, price fixing, and unfair em-ployment practices. The case dragged on for four years. Finally, in 1925, the federal court dismissed the suit, and the SPA and its mem-bers hailed the decision as a great victory and vindication of the association.[28]

Although the Southern Pine Association has never enrolled a majority of the South's lumber manufacturers, it has always had the support of large, permanent mills and the leading fig-ures in the industry. The list of presidents of the SPA reads like a who's who of lumber-men and has included many important figures among the Texas and Louisiana tycoons.[29] The association has provided a voice for the indus-try on public as well as special questions and has supported selective cutting and conserva-tion practices. The association has successfully promoted automation, greater mechanical effi-ciency, and wider uses of wood products. After more than fifty years of operation the Southern Pine Association has continued strong, active, and progressive. In it the large lumber entre-preneurs dating back to the barons of the bon-anza period have found an effective voice.

Gulf Coast lumbermen not only were active in the Southern Pine Association and other trade associations but also organized an inter-nationally known fraternal and fun society, the Concatenated Order of Hoo-Hoo. According to one source, the fraternity was formed by two lumber-industry press representatives, a news-paper man, the secretary of the Southern Lum-ber Manufacturers' Association (SLMA), a railroader, and an Arkansas lumber dealer, who were marooned together when their train was waterbound in Arkansas in 1892. Its avowed objectives are to encourage friendly re-lations, promote confidence, and foster all ac-tivities that benefit the lumber industry and humanity. Unofficially the Order of the Hoo-Hoo has been a fun-and-frolic organization given to all sorts of high jinks and merriment at the several conventions attended by its mem-bers each year. The body is governed by the Supreme Nine, of which the Snark of the Uni-verse is chairman. Other officials are Supreme Hoo-Hoo, Junior Hoo-Hoo, Scrivenotor, Bo-jum, Jabberwock, Custocation, Arcanoper, and Gurdon. In 1903 a group in Lufkin formed the Hoo-Hoo International Band. Led by Tom Humason, the band played for all sorts of gatherings as well as lumbering convocations. It helped popularize itself as well as its par-ent organization. The Concatenated Order of Hoo-Hoo grew to become an international or-ganization with several thousand members and chapters in most states and Canada.[30]

As could be expected, the lumber owners of the bonanza period took a dim view of the Pro-gressive movement. In a long editorial in 1915, the editor of the *American Lumberman* com-mented on the adverse effects of the Progres-sive agitation led by Theodore Roosevelt and other reformers. He argued that antitrust charges against even the largest operators were foolish at a time when about 48,000 sawmills were in production under unrestricted compe-tition and no single company produced as much as 1 percent of the annual cut. Owing to the conservation scare, he said, stumpage prices had doubled and tripled since 1900. Many owners had bought stumpage that they could not economically carry, and the cry for

[28] Bryant, *Lumber: Its Manufacturing and Distribution*, pp. 341–42; Southern Pine Association to Angelina County Lumber Company, December 7, 1925, ACLCo. Papers.

[29] In addition to Charles S. Keith, others mentioned else-where in this book who also served as president of SPA include John Henry Kirby, E. A. Frost, A. J. Peavy, Ernest L. Kurth, Paul Sanderson, Arthur Temple, Jr., and Thomas L. Carter. Col-lier, *The First Fifty Years of the Southern Pine Association*, p. 171.

[30] *Beaumont Enterprise*, November 6, 1955; *Lufkin Daily News* (50th Anniversary Edition), October 30, 1955; *American Lumberman*, September 12, 1908.

increased corporation taxes had caused the taxes on lumbermen's holdings to go up year after year until the owners were forced to cut the timber and dispose of the cutover acreage. The public, said the editor, was poorly informed about the lumber industry and its problems. Apparently the spokesman for the lumber operators felt that what was needed was relief from increased taxes and antitrust investigations and cooperation from the government to build up the export trade.[31] Actually the trend was in the opposite direction.

The Texas lumber operators faced these reform problems as well as special issues arising from the isolated and unusual characteristics of the industry in the state. Despite the best efforts of a series of reform governors, liberal lawmakers, and the workers themselves, the companies were able to prevent passage of any effective legislation outlawing the use of merchandise checks or coupon books in lieu of legal tender until the New Deal era. In like fashion they prevented any interference with their habitual practice of checking off deductions from the workers' pay for doctors' fees and hospitalization. Although both President Roosevelt and President Wilson were friendly to labor and labor unions grew in other parts of the United States, the Texas lumber owners, as has been noted, were completely successful in crushing attempts of organizers to form unions among the logging and mill workers in the Piney Woods. The lumbermen were also able to stave off workmen's-compensation legislation for about twenty years. Beginning about 1891, reform-minded legislators frequently introduced bills in the Texas legislature to place the responsibility and cost of industrial accidents on industry rather than the worker. The skill of the lumber lobbyists, however, together with economic and political pressure, prevented the passage of such legislation until after the election of 1912. In that year both major political candidates pledged

to enact such a law, and the national pressure brought to bear by the examples of other states was so great that opposition collapsed, and the legislature easily passed the first Texas workmen's-compensation law early in the legislative session of 1913.[32] The lumbermen continued to object to state "interference," however, and in 1917 opposed, unsuccessfully, a drive to extend and expand the coverage of the law.[33]

As might be expected, critics of the lumber industry and the operations of the lumber barons during the bonanza period were not in short supply. In 1914, during the first Wilson administration, agents for John R. Commons's United States Commission on Industrial Relations investigated representative Texas lumber operations and wrote a strongly critical report.[34] The next year a muckraking article by journalist George Creel based on the commission's report, entitled "The Feudal Towns of Texas," appeared in *Harper's Weekly*.[35] Despite these attacks few changes were made in the pattern of life or the methods of operations in the lumber industry in either Texas or Louisiana until the Great Depression and the New Deal.

Most probably neither the lumber baron nor his workers ever seriously questioned the moral or social rightness of laissez-faire. The great opportunity was here, and if he was able, he would take full advantage of it. The lumberman who emerged in East Texas and along the Sabine during the bonanza years of the early

[31] *American Lumberman*, August 7, 1915; *American Lumberman*, August 14, 1915.

[32] M. T. Jones and S. C. Allen to Alexander Gilmer, Alexander Gilmer Lumber Company Papers, University of Texas Archives, Austin, quoted in Allen, *East Texas Lumber Workers*, p. 108.

[33] *American Lumberman*, January 13, 1917. In this fight John Henry Kirby found himself confronted directly by the popular, dynamic governor James E. Ferguson. It was one of the few times during the bonanza period that the "Prince of Pines" and chief lobbyist for the lumber industry lost in a legislative contest.

[34] Peter A. Speek, "Notes on Investigations of Three Texas Lumber Towns," United States Department of Labor, Reports of the Commission on Industrial Relations, 1914 (microfilm), Forest History Collections, SFA Library.

[35] George Creel, "Feudal Towns of Texas," *Harper's Weekly* 60 (January 23, 1915): 76–78.

twentieth century was individualistic, aggressive, paternalistic, impatient of restraint, and, too often, absentee. The dire predictions of the conservationists about the destruction of the forests produced little change in his methods until the future had almost caught up with him. It is not surprising that production declined in the 1920s and that by the coming of the Depression in 1930 the history of the lumberman in Texas had turned full cycle.

In the fifty years between 1880 and 1930 the Texas lumber industry logged about 18 million acres of pine timber and cut more than 59 billion board feet of lumber. Although more than 600 sawmills of all kinds operated in Texas during most of this period, more than half of the lumber manufactured was produced by the large, permanent mills owned and operated by fewer than fifty lumbermen. A study of thirty-three major Texas lumber operators during the bonanza period revealed that only five were native Texans. Nine had come to Texas from other southern states; six had come from the Middle West, including the Great Lakes states; and seven were from the Northeast. Six were natives of foreign countries: Canada, Great Britain, and Germany. A majority could be classed as absentee owners who maintained their headquarters in some metropolitan center and only occasionally visited the logging and milling operations. At least twenty-five of the group could be described as experienced lumbermen, having engaged in the lumber business before beginning operations in Texas. Sixteen owned railroads and used them in their logging operations. At least four later extended their activities to the Pacific Coast and engaged in the lumber business there. As of 1980 only four of the lumber companies identified with the lumbermen of the bonanza period were still in existence, and only two of those were actively manufacturing lumber in Texas.

The great bonanza was a romantic, exciting period in Texas economic history, comparable to the days of the cattle kingdom and the Spindletop oil boom. Bold entrepreneurs developed great empires and amassed large fortunes. More than a generation of Texas workmen identified themselves with the near all-powerful lumber barons for whom they worked: Kirby, Temple, Carter, Kurth, Thompson, and Lutcher and Moore. At the end of the era most of the large operators moved on, leaving great problems for conservationists and the government to cope with regarding both human and natural resources.

Stumps and grass where great timber once stood.
Courtesy of U.S. Forest Service.

CHAPTER II

The Beginnings of Conservation

Trees have been my hobby for many
years.—W. Goodrich Jones

THE conservation of natural resources ordinarily is undertaken only when those resources are nearly exhausted. This truism is observed in the forests of New England, the white pines of the Great Lakes states, and the yellow-pine regions of the older states of the Southeast and later those of Texas. In time the eleventh-hour recognition of the need for conservation also came in the Pacific Northwest. In fact, Texas may be said to have become converted to the merits of conservation somewhat before the eleventh hour and thus avoided the worst consequences of the heedless destruction of its forest resources. Texas suffered severely, however, from the laissez-faire philosophy of the state and the cut-out-and-get-out attitude of many of the early lumbermen. That there were pine forests left to conserve was due to the dedicated efforts of a few civic-minded individuals, the national awareness of the urgent need for conservation awakened by Theodore Roosevelt and his Chief of the Forest Service Gifford Pinchot, and the remarkable recuperative powers of the loblolly and shortleaf species in the warm, wet climate of East Texas.

As discussed earlier, to the early Texas settler the pine forest appeared overwhelming and of little value. It stretched from the Sabine westward to the more fertile blacklands of Central Texas. In the East Texas counties the forest hindered or prevented the development of agriculture. To the lumberman and the forester, however, the pines of Texas were an impressive sight. They embraced an area of some 14 to 18 million acres—almost 30,000 square miles—stretching westward from the Louisiana border. In truth it was part of the great pine forest that covered much of the southern United States and ran out where it encountered the drier climates of the Southwest in Central Texas. One observer described its western limits in dramatic terms:

It is a striking phenomenon, this breaking up and gradual dwindling away of so vast and vigorous a forest. Not only in Texas, but far to the north, through the Indian Territory, Kansas, Nebraska, and the Dakotas, the same thing may be seen. Like a vast wave that has rolled in upon a level beach, the Atlantic forest breaks upon the dry plains—halting, creeping forward, thinning out, and finally disappearing, except where, along a river course, it pushes far inland.[1]

To many people, government officials and lumbermen alike, this vast forest seemed inexhaustible. In 1880, R. L. Sargent, of the United States Bureau of Forestry (later the Forest Service), estimated the total yellow-pine stands in Texas at 67 billion board feet and the longleaf stands of the entire Gulf South at 107 billion board feet.[2] Taking the larger figure, Henry J. Lutcher dramatized the pine resources of the South at a Senate hearing in

[1] William L. Bray, *Forest Resources of Texas*, U.S. Department of Agriculture, Bureau of Forestry Bulletin no. 47 (1904), p. 15.
[2] *Texas Almanac*, 1904, p. 145.

Washington, D.C., in 1890, stressing the potential of the southern forests:

If you will refer to Professor Sargent's forestry report for the year 1880 it will show that the amount of long leaf standing timber in those states bordering on the Gulf of Mexico was one hundred and seven billion feet. If you were to take ships at 500 tons each, load them with this one hundred and seven billion feet, place them in a direct line, stern to stern, beginning at the mouth of Sabine Pass, they would reach around the globe, and there would still be 1,600 miles of ships to come out of the Sabine Pass.[3]

Because the federal government had no public lands in Texas, no federal forestry personnel were stationed in the state, and the Bureau of Forestry had only a secondary interest in the region west of the Sabine. Nevertheless, leaders in Washington were well aware of the great potential of the Texas forests, as well as the wasteful practices of some of the early lumbermen. In 1898, Bernhard E. Fernow, chief of the Bureau of Forestry, made a trip to Texas and asked W. Goodrich Jones, a banker in Temple and a pioneer Texas conservationist, to make a survey of the region and write a report on the condition and future of forestry in the state. In his report, forwarded to Washington in 1900, Jones urged the Bureau of Forestry to cooperate with the state of Texas and the lumber industry in developing a planned cutting program that would provide for systematic replanting to prolong indefinitely the life of the forest. If such measures were not promptly taken, he warned, the great forest would disappear, and Texas would be faced with a timber famine. "Like the buffaloes," he said, "the timber is going and going fast: what escapes the big mill, is caught by the little mill, and what the little mill does not get, the tie cutters and rail splitters soon have chopped down."[4]

Jones was not the only conservationist who

W. Goodrich Jones, ca. 1883. *Courtesy of Stephen F. Austin State University, Forest History Collections.*

by the turn of the century feared the rapid destruction of the Texas forest. Writing in 1904, William L. Bray, of the Bureau of Forestry, described the Texas forest stands and the current rate of destruction and made a prediction for the future:

The longleaf pine in Texas is being cut out at the rate of some three-quarters of a billion feet of lumber each year, with a rapidly growing market and output. The ease and cheapness with which longleaf is got to the sawmill, combined with a climate which permits heavy logging throughout the year makes possible a very rapid handling of the crop. At the present rate of lumbering it would appear a reasonable estimate that the virgin pine might hold out twenty years longer.[5]

In 1904 the *Texas Almanac* estimated that there were 67 billion board feet of standing pine in Texas. In 1912 the *Almanac* reduced the estimate to 25 billion board feet and projected that the Texas lumber industries, "at

[3] Quoted in *Texas Almanac, 1911*, p. 128.
[4] Report to U.S. Bureau of Forestry, 1900, W. Goodrich Jones Papers, Stephen F. Austin State University Library, Nacogdoches; hereafter cited as SFA Library.

[5] Bray, *Forest Resources of Texas*, p. 23.

the present rate of consumption," would cease operations during the next twenty years.[6] In 1917, J. H. Foster, Texas's first state forester, compared the virgin stands and forest conditions at the time of his appointment:

The forests of longleaf pine originally formed nearly pure stands of remarkable quality and uniformity, covering extensive areas, and interrupted only by stream bottoms and moist depressions. On dry, sandy soils this species finds a refuge from the competition of other trees which it cannot endure on moister, better sites. With the long, clear stems of the longleaf pine and the open park-like and grass-covered condition of the forest floor, these forests in their natural condition were in a class by themselves among the timber regions of the world. . . . The greater part of these forests has now been cut and only where bodies of timber are held for higher values by land owning companies does one find extensive bodies of longleaf pine not in the process of being lumbered.[7]

Foster further explained that the current methods of logging with the steam skidder destroyed most of the unmerchantable timber, which earlier methods had preserved. The newer process left nothing for a second cut, and often not even enough seed trees to restock the land. Frequent fires and the notorious razorback hog killed the pine seedlings, with the result that the cutover areas were transformed into scrub-oak thickets or stump-covered grass and wastelands.[8] As an example of the rapid depletion of the forest resources of Texas, Foster estimated that, of about 563,000 acres of timber in Angelina County, only 60,000 acres of virgin trees remained plus 60,000 acres of second-growth stands. More than 300,000 acres he described as culled and cutover lands that showed "an almost total lack of longleaf pine reproduction."[9] It was indeed a gloomy picture that the young forester painted.

The rise of Gifford Pinchot on the national

forestry scene did much to stimulate conservation on state and local levels as well as in Washington. Pinchot, a graduate of Yale University, studied forestry and forest management in Germany and worked briefly as manager of the forest estate of George Vanderbilt at Biltmore, North Carolina. As the first professionally trained American forester, he emphasized many of the same concepts of utilitarian conservation that George Perkins Marsh and B. E. Fernow had advocated earlier, stressing man's responsibility to the land and to future generations. Like his predecessors, Pinchot envisioned a continuing forest and urged conservation for wise use. In 1900, Pinchot entered government service in the Division (later the Bureau) of Forestry of the United States Department of Agriculture. He soon made friends with President Theodore Roosevelt, who was also a champion of outdoor life and the preservation of natural resources. In 1905, Roosevelt appointed Pinchot chief of the reorganized and expanded Forest Service. On the advice of Pinchot the president set aside millions of acres of forest and mineral lands from the public domain as national forests, created more national parks, and designated other sites as national monuments. The two did much to promote conservation throughout the United States and to encourage state governments to take action to preserve the natural resources within their borders. To dramatize and popularize the concept of conservation, Roosevelt called a White House conference of governors in the spring of 1908 to study and discuss the problem of conservation. In 1911, to promote state participation, Congress passed the Weeks Law, which provided for matching federal funds to states that set up acceptable systems of fire protection for the forested watersheds of navigable streams and appropriated money for that purpose. This law marked the beginning of cooperation among the federal government, the states, and private industry in the protection of the forests

[6] *Texas Almanac*, 1912, p. 141.

[7] J. B. Foster, H. B. Krausz, and A. H. Leidigh, *General Survey of Texas Woodlands, including a Study of the Commercial Possibilities of Mesquite*, Texas A&M College, Department of Forestry Bulletin no. 3 (1917), p. 18.

[8] Ibid.

[9] J. H. Foster, H. B. Krausz, and George W. Johnson, *Forest*

Resources of Eastern Texas, Texas A&M College, Department of Forestry Bulletin no. 5 (1917), p. 9.

from fire and other hazards. To take advantage of this legislation, each state must set up a state conservation commission or similar agency.[10]

Among the Texas delegates to the White House Conference of Governors was W. Goodrich Jones, who was recognized by the officials of the United States Forest Service as the leading conservationist and advocate of scientific forestry in the Lone Star State. He was active in the conference and made a short speech to the assembled delegates, pledging the support of Texas under Governor Thomas M. Campbell in the cause of conservation. Jones met the president and became well acquainted with Pinchot. The two periodically exchanged letters in the years that followed.[11]

Jones was born in New York on November 11, 1860, while his parents were on a business and pleasure trip in the East. With the outbreak of the Civil War the family returned to Texas in 1861, and young Jones spent the war years in Galveston and Houston. After studying in New York, in 1873 he accompanied his family to Europe, where he attended a German grammar school. There he studied forest science and went on a walking tour through the Black Forest with his father. He talked with the rangers and village managers who made their livings from the forest, as their ancestors had before them. There Jones developed a deep appreciation of the beauty and commercial value of a well-managed forest. He studied the care with which the German foresters planted, cut, and replanted, producing a continuous harvest of forest products. He accepted their maxim that when a lumberman cuts a tree from the forest he must plant another to replace it.[12]

After the family returned from Europe, Jones prepared for college and entered Princeton University, from which he graduated in 1883 with a degree in business. He returned to

Texas, and in 1888, after serving an apprenticeship in banking in Galveston and other South Texas towns, he became president of a bank in Temple in Central Texas. There he had time and resources to promote his favorite hobby, tree planting and the conservation of the forests.[13]

"My first impressions of Temple were unfavorable," wrote Jones at a later date, "as not a tree was to be seen." The need to have trees around him so possessed him that he planted pecan trees in large tin cans, which he placed on his hotel windowsill. As he said, "This was the town's first tree planting." He soon interested others in planting trees, and young seedlings of many varieties began to appear throughout the city. Jones was nicknamed the "Tree Crank," but he continued his crusade with increasing success.[14]

Shortly afterward, Jones played a leading part in the designation of February 22 as Texas Arbor Day. Following a suggestion of Governor Sul Ross that Texas, like many other states, should have an Arbor Day, Jones and other citizens of Temple passed a resolution setting aside Washington's Birthday as Arbor Day and urging the legislature to so designate it. With the aid of the local state representative and the governor, such a bill was unanimously passed by the legislature and signed by the governor on February 22, 1889. Within a year the Texas Arbor Day and Forestry Association was established, with Governor Ross as president and Jones as secretary. Its purpose was

to encourage tree planting on Arbor Day in the State, the conservation, management and renewal of forests, the collection of forest statistics and the advancement of educational and legislative tree knowledge. . . . It shall especially endeavor to diffuse the knowledge thus gained and create a wide spread interest in the subject.[15]

---

[10] Gifford Pinchot, *Breaking New Ground*, pp. 347–55; Henry Clepper, "Crusade for Conservation," *American Forests*, October, 1975, chap. 5.

[11] Newton C. Blanchard et al., eds., *Proceedings of a Conference of Governors in the White House, Washington, D.C., May 13–15, 1908*, pp. xix–xxxi, 190–91.

[12] Anna Jones, "Early Days of W. Goodrich Jones," W. Goodrich Jones Papers.

[13] "Life and Work of W. Goodrich Jones," speech at dedication of W. Goodrich Jones State Forest, Conroe, Texas, May 19 1949, W. Goodrich Jones Papers; *Gulf Coast Lumberman* 38 (August 15, 1959): 18.

[14] "Life and Work of W. Goodrich Jones," W. Goodrich Jones Papers.

[15] W. Goodrich Jones, "Texas Arbor Day and Forestry Association Constitution," W. Goodrich Jones Papers.

One effect of the White House conference was to stimulate conservation activity in
the states. By then the Texas Arbor Day and
Forestry Association had become inactive. A
new society, the Conservation Association of
Texas, was organized and called a meeting of
a Conservation Congress at Fort Worth, in
April, 1910. At the congress representatives of
state government and the lumber industry as
well as professional conservationists and concerned citizens heard speeches by a number
of prominent figures, including a representative from the United States Forest Service,
J. C. Gips, and J. Lewis Thompson, president
of the Yellow Pine Association.[16] The association adopted a constitution and bylaws stating
as its objective "the development and conservation of all the natural resources of this state."
The congress also proposed that the state of
Texas create a state agency to supervise conservation work, either a "commissioner of conservation" or "a state department of forestry."
At this meeting of the Conservation Congress,
Jones served as secretary, and Edward R. Kone,
state commissioner of agriculture, served as
president.[17]

The idea of a state department of forestry
had been proposed to Jones by B. E. Fernow,
of the United States Division of Forestry, back
in 1898 at the time of Jones's East Texas survey. Increasingly, especially after the White
House conference, Jones directed his energies
toward the establishment of such a department
as the best means to promote his goal of conservation and reforestation. Unfortunately, despite its well-publicized beginning, the Con-

servation Association of Texas soon languished and by 1914 was virtually defunct.
Undaunted, Jones again sought to form a voluntary organization that would draw the mass
membership and broad support needed to
prod the legislature into effective action.[18]

Early in November, 1914, Jones called a
meeting of about twenty conservation-minded
citizens at Carnegie Library in Temple. This
group organized the Texas Forestry Association with the avowed objective of creating a
state department of forestry and the development of a comprehensive forest-conservation
program. Jones became president of this
group. Jack Dionne, editor of the *Gulf Coast
Lumberman*, was elected secretary, and that
periodical was named the association's official
organ. The officers at once took steps to en-
large the membership to statewide proportions. They set dues at a nominal twenty-five
cents for annual memberships, five dollars for
patron members, and ten dollars for life members. The group also drew up a model bill
creating a department of forestry and designated Jones to lead a delegation to Austin to
see to the bill's introduction and passage. Thus
Jones became the leading lobbyist for creation
of a forestry department and the employment
of a state forester, both of which would require
a considerable legislative appropriation.[19]

Jones and his colleagues prepared for the
coming legislative session with care, mustering
all the support they could gather. Jones had
written to the United States Forest Service explaining what the Texas Forestry Association
was seeking and asking for assistance. To help
in the preparation and presentation of the bill
came J. Girvin Peters, chief of the State Cooperation Division of the Forest Service.[20] Once

[16] *Dallas Morning News*, April 7, 1910. J. Lewis Thompson
was among the first of the major Texas lumbermen to take an
active part in the conservation of forest resources. In a speech,
"Attitude of Lumbermen Toward Forestry," delivered in 1925,
Thompson argued that the great virgin forests were all but depleted and that fires and razorback hogs would prevent natural
reseeding. He called on lumbermen to support the state and the
Conservation Association in promoting reforestation and forest
protection. *American Lumberman's Review* 48, no. 8 (April 25,
1925): 30.

[17] "Texas Conservation Congress Opens," *Dallas Morning
News*, April 6, 1910; memoranda, W. Goodrich Jones Papers.

[18] "Life and Work of W. Goodrich Jones," W. Goodrich Jones
Papers.

[19] W. Goodrich Jones, "The Dawn of Texas Forestry Conservation," address to Texas Forestry Association, May, 1947,
Jones Papers.

[20] P. S. Ridsdall to Jones, December 18, 1914, Jones Papers.
The Texas Forestry Association was committed to pay Peters's
expenses, amounting to about $350. The TFA raised the
money through contributions by its members.

the bill was in final form, Jones and Peters visited several major cities and lumber-company headquarters, soliciting support for the measure. The officials at Texas A&M College were enthusiastic about the project. President William Bennett Bizzell and Dean E. J. Kyle volunteered to go to Austin to talk to legislators about the bill. At the University of Texas J. T. Phillips, professor of geology, also agreed to lobby in behalf of the forestry program. Lumberman J. Lewis Thompson favored the bill but declined to take an active role as sponsor, saying that he was out of town too often and too busy with his own affairs. John Henry Kirby assured Jones that he supported the bill but thought that the measure would fare better if he stayed in the background.[21] United States Senator Morris Sheppard, who had attended the White House conference, offered his support and arranged to have a report by Peters, "A Forest Policy for Texas," published in all the state's leading newspapers.[22] Jones had yet one more ally in his fight for the establishment of a department of forestry: the new governor, James E. Ferguson, was a fellow citizen of Temple and a personal friend. Ferguson had progressive leanings and favored the conservation program. At Jones's urging, he had included a paragraph on Texas forestry needs in his inaugural address.[23]

Despite the advanced planning, the forestry bill faced a hard struggle and considerable amending before it finally became law. Richard F. Burges of El Paso, a member of the TFA who later served as president of the association, agreed to introduce the bill and act as its sponsor. None of the members of the Texas House Committee on Forestry, to which the

bill was referred, were conservationists or members of the FTA. They were slow to act, and more than a month passed before the bill was reported, after considerable prodding by Jones, the governor, and others, with the recommendation for passage with a number of amendments.[24]

By this time it was apparent that the lawmakers were in no mood to establish a new department or independent agency to administer the conservation program. As an alternative Bizzell and Kyle proposed that the department be set up on the Texas A&M campus and administered by the Board of Directors of the college. Some supporters of the measure, including lumberman Thompson, were opposed to linking it with Texas A&M, but Jones and Peters had no major objections and hoped that the proposal would end most of the opposition to the bill. On the motion of Burges this and other amendments were incorporated into the bill, and after several close test votes the bill passed the house in mid-March by a margin of seven votes and was sent to the senate.[25]

In the upper house the conflict was sharp but quickly resolved. After further amendments and clarification, the bill survived the crucial test vote by the margin of one vote. Evidently Jones and his fellow lobbyist had carefully lined up support in the senate while the bill was still being considered in the house. The conference committee quickly agreed to the senate changes, and the approved measure was sent to the governor for his signature on March 20, 1915.[26]

Thinking that their work was done and that the forestry department bill was law at last, Jones, Peters, and Kyle left Austin and returned to their homes. President Bizzell, however,

[21] Jones, "The Dawn of Texas Forest Conservation"; J. Lewis Thompson to Jones, December 14, 1914; John Henry Kirby to Jones, November 30, 1914, W. Goodrich Jones Papers.

[22] Morris Sheppard to Jones, December 5, 1914, W. Goodrich Jones Papers. Peters's report, "A Forest Policy for Texas," is reprinted in James W. Martin, "A History of Forest Conservation in Texas, 1900 to 1935" (master's thesis, Stephen F. Austin State University, Nacogdoches, 1966), appendix D.

[23] Jones, "The Dawn of Texas Forestry Conservation," W. Goodrich Jones Papers.

[24] Ibid.; *House Journal of the Texas Legislature*, 34 Leg., reg. sess., 1915, p. 109.

[25] Both Bizzell and Kyle had earlier proposed such an alliance, pointing out that both the state chemist and the state entomologist had headquarters at Texas A&M. Kyle to Peters, November 11, 1914, Jones Papers; *House Journal*, 34 Leg., reg. sess., 1915, pp. 908–909, 1080.

[26] *Senate Journal of the Texas Legislature*, 34 Leg., reg. sess., 1915, pp. 38, 1221, 1230.

remained in Austin to conduct some private business and to see the conservation measure through. It is fortunate that he did. After several days went by and the governor had not yet signed the measure, it became apparent that members of the opposition had persuaded him that a salary of $3,000 a year was too high for the state forester and that some A&M graduate would be glad to take the job for half that amount. The total appropriation of $10,000 was said to be too high and wasteful. Bizzell at once recalled Jones, Kyle, and Phillips, and the four met with Governor Ferguson. After Jones had described the professional qualifications of a state forester and explained the need to establish the forestry department without further delay, Ferguson agreed to sign the measure. Jones quickly lent him his own pen, and the governor signed the act into law. Thus the Texas State Forestry Department was established.[27]

Although the authorization had been made and the appropriation had been approved, the department still had to be organized and made to function. One of the first tasks was to find a "technically trained forester with no less than two years' experience" to inaugurate the program at the salary offered. In the June, 1915, issue of the *Gulf Coast Lumberman*, Jones placed an advertisement, hoping to find a suitable applicant:

Help Wanted
The applicant must be a "Chesterfield," an orator, a lecturer, a mixer, a highly trained specialist in the theory of forestry and withal a practical woodcraftsman. Salary $3000 annually.[28]

In answer to this somewhat whimsical ad, J. H. Foster, of Vermont, applied and was ap-

pointed to the position. Foster held the degree of master of forestry from Yale (1907) and had had several years' experience as a forester in his native New England. One of his first acts was to apply to the federal government for matching funds for fire protection under the Weeks Law. He organized the first fire-protection system in the state, and within a year the new Department of Forestry had about 7.5 million acres under fire protection. Six fire patrolmen, with headquarters at Lufkin, Longview, Livingston, Linden, Jasper, and Tenaha, rode through their districts and talked with the citizens about fire protection. In an effort to educate the public regarding fire prevention and forest management, Foster gave twenty-five lectures before lumbermen's and stock raisers' groups, women's clubs, and teachers' organizations. He prepared a display for the state fair in Dallas, and he wrote fifteen hundred letters on forest-related subjects.[29]

At the same time Foster wore at least three other hats. He served as secretary-treasurer of the TFA, carried the designation of chief of the Division of Forestry in the Texas Agricultural Experiment Station, and was professor of forestry in the teaching division of Texas A&M. Beginning in the fall of 1916, Foster taught six courses in forestry and conservation for agriculture majors in the college. In addition to these duties Foster found time to write and edit six bulletins on forest conditions in Texas. In his second annual report he described what he had done and asked the legislature to appropriate no less than $20,000 to advance the work thus begun.[30]

Not everyone was pleased or impressed by

[27] Jones, "The Dawn of Texas Forestry Conservation"; Bizzell to Jones, March 22, 1915, W. Goodrich Jones Papers. The Jones pen used by Governor Ferguson to sign the original forestry bill is on exhibit in the Texas Forest Service headquarters, College Station.

[28] *Gulf Coast Lumberman*, June, 1915; Edward R. Wagoner, "The Green Gold of Texas," *Texas Forestry Association 50th Anniversary*, reprinted from *American Forests*, September, 1964.

[29] J. H. Foster, H. B. Krausz, and George W. Johnson, *First Annual Report of the State Forester*, Texas A&M College, Department of Forestry Bulletin no. 4 (1916), pp. 1–7; John A. Haislet, "Texans Evolve a State Forestry Agency," *Texas Forests and Texans*, May–June, 1964, p. 6.

[30] J. H. Foster, *Second Annual Report of the State Forester*, Texas A&M College, Department of Forestry Bulletin no. 8 (1917), p. 7; Robert S. Maxwell and James W. Martin, *A Short History of Forest Conservation in Texas, 1880–1940*, Stephen F. Austin State University, School of Forestry Bulletin no. 20 (January, 1970), p. 29.

the work that Foster had begun in the first two years. Many legislators considered the forestry department a fad and that the $10,000 provided for its establishment had been wasted. Indeed, when the legislature met in 1917, the House Committee on Appropriations passed a resolution to cease allocating funds to the Department of Forestry on the grounds that the agency was useless and the expenditure of funds was of no benefit. Again President Bizzell was the first to hear of this action, and he promptly called Jones and Phillips to College Station, where, with Dean Kyle, they swung into action. As Jones later said, "We then made the wires hot, in sending telegrams to all the leading newspapers in Texas, asking them for their cooperation in restoring the appropriations."[31] Jones sent a personal letter to the editor of the *Houston Post* asking for an editorial in support of the forestry program. He again asked John Henry Kirby to appear with him before the house and senate committees in behalf of the Department of Forestry. Kirby again declined but assured Jones that he was entirely in support of the conservation program. He remarked that if he came out strongly in favor of the project the legislators would probably think that he was after some selfish gains, since they already thought that he owned half of East Texas. He promised, however, to work for the measure quietly in the background.[32]

Within a few weeks Jones and his colleagues had secured the endorsement of the Texas State Teachers' Association, the Texas Farmers' Institute, the Farmers' Congress, the Texas Cattle Growers' Association, the General Managers' Association of Texas Railways, and others. When these organizations, influential state newspapers, and Jones's own persuasive arguments converged upon the lawmakers, the opposition wilted. The final appropriations bill restored the Department of Forestry to the list and raised its budget to $12,000. Although still short of the $20,000 that Foster had requested, the work of Jones, Bizzell, and their fellow members of the TFA had ensured the survival of the program and had won a small increase in funds. The principle of a state-supported conservation agency was secure.[33]

At the end of 1917, Foster resigned, and the Texas A&M board of directors had to find a successor. As the first Texas state forester and director of the Department of Forestry, Foster had done well in setting up the program and inaugurating the fire-protection service. But Foster was not happy in Texas. He was overworked, with a multitude of duties, and he was discouraged by the almost successful efforts of the 1917 legislature to destroy the conservation program. Foster returned to his native New England, where he served as state forester of Vermont for more than thirty years.[34]

The second Texas state forester was E. O. Siecke. Born and educated in Nebraska, Siecke had worked under Pinchot in the United States Forest Service. In 1910 he served as an assistant professor in the Oregon College of Agriculture, and the next year he became deputy state forester of Oregon, where he remained until 1918. As Texas state forester Siecke worked well with W. Goodrich Jones, and the two became warm friends. He also won the respect of the Texas A&M administration and the state legislature for his ability, his devotion to the cause of conservation, and his staunch character. It seemed a happy case of the man meeting the job, for Siecke remained head of the Texas conservation program for twenty-five years, until he retired in 1942.

Not without humor, Siecke enjoyed telling the story of his wife's introduction to Texas. After he had accepted the position, his wife and family traveled by train from Oregon and

[31] Jones, "The Dawn of Texas Forestry Conservation," p. 3, W. Goodrich Jones Papers; Maxwell and Martin, *A Short History of Forest Conservation in Texas*, pp. 29–30.

[32] Jones to H. T. Warner, April 23, 1917; Kirby to Jones, May 14, 1917, W. Goodrich Jones Papers.

[33] "Jones to the Senators and Representatives of the Thirty-fifth Legislature," February 19, 1917; Jones, "Dawn of Texas Forestry Conservation," pp. 3–4, W. Goodrich Jones Papers.

[34] Haislet, "Texans Evolve a State Forest Agency," p. 6.

E. O. Siecke. *Courtesy of Texas Forest Service.*

California to join him. Riding east of El Paso, Mrs. Siecke looked out at the treeless, West Texas plains. A stranger in the next seat politely inquired about her destination and her husband's occupation. She replied that he was the new state forester of Texas. The amazed stranger gazed out the window at the sagebrush and cactus and then remarked, "Well, lady, all I can say is that Texas doesn't need a State Forester or else they need one awfully bad!"[35] Texas needed a state forester "awfully bad," and they had found him.

Noting that his predecessor's time had been largely taken up with various teaching duties and speaking engagements, which left him little time to promote the work of the Forestry Department, Siecke sought to free his time for the more important tasks. In 1921, Representative Gary Standford of Timpson, a member of the House Appropriations Committee, attached a rider to the general appropriations

bill prohibiting the officials at Texas A&M from requiring the state forester to teach classes in the college. Years later Siecke related that the initiative for this rider had come from the state forester himself, that he had proposed and largely supplied the language of the rider to his friend Standford.[36]

Under Siecke the Texas Forestry Department progressed and expanded. More land was placed under fire protection, more patrolmen were employed, fire towers were added, and by 1925 an assistant forester had become necessary. The department's budget rose from $12,000 in 1918, when Siecke arrived, to $52,000 in 1924–25, and to $64,000 in 1932. In addition, a more liberal revision of the Weeks Law, the Clarke-McNary Law of 1924, provided another $40,000 in federal matching funds for fire-control work. The new law stimulated the participation of private owners in fire-prevention work, and "protection units" were formed. Progressive landowners, led by Arthur Temple and others, contributed two cents per acre of timber holdings, which augmented state and federal funds to provide intensive protection for these areas.[37]

The year 1926 was a major milestone in the development of forest conservation in Texas. In that year the Texas A&M Board of Directors recognized the agency as one of the four major divisions of the college and changed the name of the department to Texas Forest Service. Siecke was given the title director of the Texas Forest Service.[38] Under Siecke the Texas Forest Service developed into a strong, efficient, professional organization relatively free of political interference. Indeed, during the first sixty-five years of its existence, the TFS has had only five directors. This record has

[35] Conversation with E. O. Siecke, Galveston, May 23, 1967, Oral History Collections, Forest History Collections, SFA Library; Wagoner, "The Green Gold of Texas," p. 4; Haislet, "Texans Evolve a State Forest Agency," p. 6.

[36] Conversation with E. O. Siecke, Galveston, May 23, 1967, SFA Library.

[37] Haislet, "Texans Evolve a State Forest Agency," p. 6; "Fortieth Anniversary of the Texas Forest Service," *Texas Forest News* 34, no. 2 (March–April, 1955): 4–7; Maxwell and Martin, *A Short History of Forest Conservation in Texas,* p. 33. Under the fire-protection program fire losses in Texas declined from 1,131,500 acres in 1916 to 244,536 acres in 1930.

[38] Haislet, "Texans Evolve a State Forestry Agency," p. 6.

A tree cab, the forerunner of the lookout tower.
Courtesy of Texas Forest Service.

Forester checking with headquarters. *Courtesy of Texas Forest Service.*

Texas Forest Service firefighter with tools. *Courtesy of Texas Forest Service.*

been in marked contrast to that of many of its neighbors.[39]

During the Siecke years Texas made progress in other areas of forest conservation. The state acquired its first state forest in 1924, when 1,720 acres of cutover land were acquired in Newton County; this holding was named the E. O. Siecke State Forest in 1951. The next year the Texas Prison Board transferred to the Department of Forestry 2,630 acres in Cherokee County, which became the I. D. Fairchild State Forest. In 1927 the state purchased 1,725 acres in Montgomery County, which was designated the W. Goodrich Jones State Forest in 1949. Also in 1927, John Henry Kirby donated 600 acres in Tyler County for the John Henry Kirby State Forest. These formed the nucleus of the state-forest system. At the same time the Texas Forest Service took steps to promote reforestation. Jones, who had been calling for a comprehensive tree-planting campaign since the turn of the century, was now

joined by Siecke and Governor Pat Neff to implement the efforts of private citizens. In a project to provide seedlings at cost for reforestation, the Texas Forest Service developed a tree nursery on the land of the future E. O. Siecke State Forest in 1926 and another in 1928 on what would become the W. Goodrich Jones State Forest. By 1930 the two nurseries were producing one million seedlings a year. The availability of pine seedlings greatly encouraged lumbermen with an eye to the long-range future, such as Temple, Kurth, Carter, and Kirby, to reforest systematically parts of their cutover acreage. Although the state did not pass legislation requiring the lumber owners to plant a tree when they cut one, the Texas Forest Service's campaign of education, fire protection, and reforestation did much to produce a vigorous second-growth pine forest in East Texas within one generation after the bonanza era ended.[40]

[39] *Texas Forestry*, April, 1974, pp. 38–39.

[40] Jones, "Speech at Dedication of the W. Goodrich Jones State Forest," May 19, 1949, W. Goodrich Jones Papers; Max-

The names of the men who promoted the cause of conservation in Texas read like a roll of honor of Texas forestry. In addition to serving as state forester and director of the Texas Forest Service from 1918 to 1942, Siecke was secretary of the TFA from 1918 to 1936. He was also the first chairman of the Gulf States Section of the Society of American Foresters and a member of the editorial staff of the *Journal of Forestry*. After his retirement the TFA elected him president emeritus of the association. Among governors, in addition to James Ferguson and Pat Neff, Dan Moody and James Allred were knowledgeable about forestry matters and showed a genuine concern for the future of conservation. Among prominent lumbering figures who served as presidents of the Texas Forestry Association have been Ernest L. Kurth, R. W. Wier, Paul Sanderson, H. W. Whited, Hoxie H. Thompson, N. D. Canterbury, S. W. Henderson, Jr., and R. W. Wortham, Jr.[41]

In any study of the beginnings of conservation in Texas, one must go back to W. Goodrich Jones. It was he who translated Theodore Roosevelt's and Gifford Pinchot's call for the protection and renewal of natural resources into a practical program for Texas. He used the enthusiasm generated by the White House governors' conference to found the Texas Forestry Association and to nurture it during its fledgling years. With the TFA as a base, Jones was instrumental in persuading the legislature to establish the Texas Forest Service and to give it increasingly generous support as its programs expanded. An urban banker and businessman, Jones never pretended to possess the scientific training of the professional forester; rather he was the enlightened layman who was keenly aware of the problems and the need for conservation in Texas. He tirelessly sought to bring together leaders of the timber industry, state officials, professional foresters, and members of the general public to identify and solve

W. Goodrich Jones, ca. 1926. *Courtesy of Stephen F. Austin State University, Forest History Collections.*

these problems. Jones was not a "preservationist" who would lock up the forests and leave them untouched. He was more a "utilitarian," who, like Pinchot, promoted conservation for wise use.

Jones was concerned about the broader aspects of conservation—soil, rainfall, wildlife, flowers, and grasses, as well as trees. He constantly promoted the establishment of parks and forests by wealthy citizens, cities, the state, and the national government. He wanted every town to be a "green town," with streets lined with well-chosen trees and a convenient "commons" to provide rest and recreation areas for its citizens. Thus it seems proper that his name has been perpetuated by the W. Goodrich Jones State Forest in Montgomery County and that he has been acclaimed the "Father of Forestry in Texas."[42]

well and Martin, *A Short History of Forest Conservation in Texas*, pp. 34–35.

[41] Conversation with E. O. Siecke, Galveston, May 23, 1967, SFA Library; *Texas Forests and Texans*, May–June, 1964, p. 4.

[42] For a more detailed account of the career of W. Goodrich Jones, see Robert S. Maxwell, "One Man's Legacy: W. Goodrich Jones and Texas Conservation," *Southwestern Historical Quarterly* 77 (January, 1944): 355–80.

CHAPTER 12

The Bridge of Ships

Ships will win the War,—We must build a
bridge of ships across the sea.
—George Creel

EVER since the days of the "broad arrow" in colonial America the majestic pine tree was a valued and important commodity for export and shipbuilding.[1] In the South builders and shipwrights especially prized the longleaf pine for its straight length, superior strength, and ease of working. Not only was pine valuable for masts and spars, but builders used it for structural timbers, siding, decks, and a thousand other purposes. In wood-scarce England and western Europe there was a steady market for timbers, dimension material (boards larger than 1 by 12 inches), other boards, poles, and pilings. The ships to carry this lumber as well as other products of the New World were often built in America—almost totally the products of the pine forests.

Along the Texas Gulf Coast early sawmill operators experienced a similar demand. Despite a shortage of lumber of all kinds in Central Texas, reports from as early as the 1850s tell of increasing exports of lumber and lumber products. As described earlier, most of the first commercial sawmills in Texas were set up on navigable rivers on or near the Gulf Coast. When the local timber supply became exhausted, the lumbermen extended their log-

ging operations as far upriver as necessary and floated the cut logs downstream to the mill. The sawed lumber could conveniently be loaded on ships for coastwise shipment or export.[2] It was not surprising that Galveston, Houston, Beaumont, and Orange became leaders in the export trade as well as in the manufacture of lumber. As early as 1860 it was reported that 1 million board feet of sawed lumber, about 12 million shingles, and 97,000 staves had been exported from Sabine Pass that year.[3]

Even earlier, Robert Harris was reported shipping yellow-pine and oak lumber and timber from Buffalo Bayou to the Mexican ports of Tampico and Matamoros.[4] By 1880 shipyards were operating in Galveston Bay, Indianola, Matagorda, Sabine Pass, and Orange. Shipwrights constructed sloops, schooners, tugs, and steamboats, using the best longleaf timber available. Most were designed to carry cargoes of lumber and lumber products.[5]

Each of the Texas ports was hampered by a sandbar running along the coast that in places allowed clearance of as little as four to six feet. The bar had to be dredged to permit the entrance of larger oceangoing vessels. Before

[1] In the late seventeenth century the British crown directed the "Surveyor General of His Majesty's Woods in North America" to go through the woods at convenient times and mark all "mast trees" with a "broad arrow" and thus reserve them for his majesty's navy. Richard G. Lillard, *The Great Forest*, pp. 122–37.

[2] Vera Lea Dugas, "Texas Industry, 1860–1880," *Southwestern Historical Review* 49, no. 2 (October, 1955): 154–82.
[3] *Texas Almanac, 1860.*
[4] Cleo F. Evans, "Transportation in Early Texas" (master's thesis, St. Mary's University, San Antonio, 1940), pp. 76–85.
[5] Dugas, "Texas Industry, 1860–1880," p. 173.

dredging operations began, the shippers would load the lumber or other products on a barge and tow it out to a ship waiting in the Gulf. Soon after his arrival in Texas, Henry J. Lutcher became interested in the export trade and in the improvement of harbor facilities at Sabine Pass. In 1888 he loaded his first vessel of substantial size with crossties bound for the Mexican railroad at Tampico. He loaded the ties on a barge, which was towed by tug from Orange down the Sabine to the *Comet*, a schooner lying at anchor in the harbor.[6] In 1890, Lutcher appeared before a congressional committee on rivers and harbors urging Congress to appropriate funds to convert Sabine Pass into a permanent deepwater port. As a result appropriations were made to improve port and channel conditions on a regular schedule. According to one lumber historian, "Lutcher did more to promote the interests of Sabine Pass and to open up that port to deep water than any other individual in the South or Southwest."[7]

Exports of lumber continued to grow after the turn of the century. In 1902 the *American Lumberman* noted that during the past year exports of lumber and timber with a total value of more than $8 million had been shipped from Galveston to such foreign markets as Belgium, the Netherlands, Denmark, Germany, France, England, and Mexico.[8] The next year Lutcher and Moore announced that the company would charter two ships, the *Alps* and the *Clover*, to carry a cargo of yellow pine to Veracruz, Mexico. John Henry Kirby reported having made three shipments aggregating 5 million board feet to Liverpool. "The prices offered," he added, were "good." The same year Kirby shipped a cargo of prime

longleaf lumber to Rotterdam, transporting it by barge from Beaumont downriver to Port Arthur, where it was transloaded on the ship.[9]

For the lumbermen of Texas, particularly those of Beaumont and Orange, their years of agitation and lobbying reached a climax in 1911, when Congress voted funds to convert the Sabine and lower Neches into a deepwater port.[10] The project provided for twenty-six-foot channels, protected by saltwater guard locks, which would allow larger vessels to anchor at pierside in Beaumont and Orange. The project was completed and opened for traffic early in 1916.[11]

Traffic in lumber increased still more with the new facilities. In May, 1916, the British steamer *Westlands* docked and loaded directly at pierside in Orange, the first vessel of that size to use the new deepwater port. When it sailed, it carried more than 2.5 million board feet of sawed lumber bound for London. A few weeks later the British steamship *Pikepool* sailed from Beaumont with about 3 million board feet of lumber. Other shipments followed, and by year's end more than 100 million board feet of yellow pine (about 5 percent of the total cut for that year) had been exported through Texas Gulf ports. At the same time other facilities had improved. Joseph Weaver and his son had established a shipyard and dry dock at Orange about 1899. When larger ships began using the harbor facilities, the Weavers put in modern marine ways and hoists so that they could handle the largest vessels entering the Sabine. In 1916 the Standard Export Lumber Company, perhaps the largest exporter of southern pine, established offices in Orange, Beaumont, and Galveston. Henry Piaggio, representing London and Italian exporters and steamship owners, also set up offices in Texas

[6] Lutcher and Moore Papers, Home Office, Orange; Marcus E. Sperry, *The Port of Orange, Texas, U.S.A.*, copy in Forest History Collections, Stephen F. Austin State University Library, Nacogdoches; hereafter cited as SFA Library.

[7] James E. Defebaugh, *American Lumbermen*, 2:380. Congressman Samuel B. Cooper and Senator Joseph W. Bailey made significant contributions to the success of this project. Sperry, *The Port of Orange*.

[8] *American Lumberman*, December 27, 1902.

[9] *American Lumberman*, February 14, 1903, p. 47; March 7, 1903, p. 35; August 8, 1903, p. 35.

[10] These were matching funds equaling monies raised by Orange and Jefferson counties, which had created "navigation districts" for that purpose. Sperry, *The Port of Orange*.

[11] Ibid.; *American Lumberman*, May 20, 1916.

Schooner under construction at Beaumont. *Courtesy of* Gulf Coast Lumberman.

to expedite lumber shipments from the Orange-Beaumont area.[12]

As might be expected, the larger companies dominated the export trade, and Lutcher and Moore dominated the trade at the new Port of Orange. Long an advocate of a steady export trade as a hedge against the fluctuations of the domestic market, Lutcher regularly extended his overseas contacts and improved his facilities for handling the trade. Possessing some of the finest longleaf stands in Louisiana and Texas, Lutcher proudly claimed that his export trademark, "Lutcher-Orange," was known the world over for first-quality lumber. Because of the high freight rates charged by contract shipping lines, Lutcher (and later William H. Stark) built up his own fleet, and by 1916 the company owned such vessels as the *Roseway*, the *Doane*, the *Hugo*, the *San Ramon*, the

Cuba, the *Sabine*, and the *William H. Murphy*. He also kept a number of ships under lease.[13] Lutcher was not, of course, the only Texas lumber tycoon who engaged in the export trade. John Henry Kirby operated from both Beaumont and Houston under several company names. The Miller-Link Company, of Orange, also had an extensive export business, as did Long-Bell, Alexander Gilmer, and others.

Perhaps the most exciting event of 1916 for persons interested in the Texas lumber industry was the building and launching of the schooner the *City of Orange*. A number of ships had been constructed in the Gulf ports before then, some of considerable size. The *City of Orange*, however, was hailed as the largest ship built on the Gulf Coast up to that time. It was largely constructed of longleaf

[12] *American Lumberman*, May 20, 1916, p. 30; July 8, 1916, p. 53; September 9, 1916, p. 51; December 23, 1916, p. 51; Sperry, *The Port of Orange.*

[13] *American Lumberman*, August 14, 1915, p. 58; July 22, 1916, p. 62; September 23, 1916, p. 43; November 18, 1916, p. 54; Sperry, *The Port of Orange.*

The *City of Orange* just before its launching. *Courtesy of* Gulf Coast Lumberman.

yellow pine. Built by Captain Fred Swailes and Company, of Orange, it was laid down in the summer of 1916 and launched on November 20 of the same year. It was described as 250 feet long and 43 feet wide with a 23-foot-deep hold. More than 1 million board feet of choice longleaf pine were worked into the ship, as well as Oregon fir for spars and liveoak for the bow and stern. When completed, the *City of Orange* was rigged as a five-masted schooner with two auxiliary 100-horsepower diesel engines to provide additional power. Built for Henry Piaggio and Company, the vessel was designed for the lumber trade and was expected to have a capacity of about 1 million board feet of lumber. When the big schooner was launched, it was christened in elaborate ceremonies and regarded as a signal achievement. Perhaps it was, for it was a foretaste of things to come.[14]

The outbreak of World War I in August, 1914, brought temporary disquiet and dislocation to the Texas lumber industry. Because of wartime uncertainties many exporters curtailed or canceled their orders, and manufacturers suddenly found themselves without contracts and with excessive stocks on hand. Some companies cut production, laid off some of their workers, or even closed down their mills. The situation was, of course, only temporary, but the editor of the *Gulf Coast Lumberman* accurately summed up the initial reaction:

The immediate, almost instantaneous, effect of the war, was the receipt of cables from Europe, stopping the shipment of lumber absolutely. Every export lumber or timber contract bears a cancellation clause in case of war. The outbreak of hostilities found many vessels laden with yellow pine from Southern ports on the high seas headed for Europe. None have been heard from as we go to press. It found many vessels loading at Gulf ports with lumber and timber for war-laden Europe. Every operation was stopped. Half laden vessels were unloaded, work was immediately stopped.[15]

The market depression did not last long in Texas, however. The total production of lumber, which had declined to 1.5 billion board

[14] *Gulf Coast Lumberman*, January 1, 1917, pp. 26–29; *American Lumberman*, November 5, 1916, pp. 56–57; November 25, 1916, p. 20.

[15] "European War and the Lumber Situation," *Gulf Coast Lumberman*, August 15, 1914, pp. 4–5.

feet in 1914, rose to 1.75 billion board feet in 1915 and still higher in 1916. The Allies, who controlled the Atlantic sea-lanes, soon found that the war greatly increased their lumber and timber requirements, ranging from boxes and crates to temporary houses, army-camp structures, and timbers to shore up trenches and dugouts. The prime source of lumber was the southern United States, particularly the Gulf Coast.

The German submarine campaign, the Central Powers' chief weapon against the Allied blockade, acutely pointed up the scarcity of ships sailing under the American flag. As British and French purchases increased, goods piled up on American docks, and needed imports failed to arrive. A cry rose from the press and the public for a reactivation of shipbuilding yards. In fact, any vessel that floated was in demand. The *Gulf Coast Lumberman* described the shift from the pessimistic days of 1914 to the boom of 1916:

Old shipyards idle for years have been bustling with activity for more than two years. New yards have sprung up everywhere along the American seaboards and today all plants are straining every resource to meet the orders with which they have been flooded.

Plants along the Texas and Louisiana coasts are no exception. After more than twenty years of idleness so far as the construction of seagoing vessels is concerned, they have come back. They are building vessels to assist in carrying the commerce of the world.

For building wooden vessels, the yards on the East Texas and West Louisiana coasts are almost ideally situated. They have an abundance of timber from which to select materials and a climate that guarantees maximum efficiency from a working force throughout the year.[16]

In the effort to build as many ships as possible, many different types of construction were tried—steel, wooden, and even concrete ships. To those who contended that wooden ships would be ineffective and a waste of time and

money, advocates replied that there were many good reasons for wooden-ship construction. They could be built much more quickly than steel ships, a great timber supply was at hand ready for use, and, indeed, they might prove able to withstand a submarine torpedo as well or better than a steel ship. One enthusiast claimed, "One torpedo, well-aimed, can sink a great steel ship of 10,000 tons—but as much effort and explosives are required to destroy the modest 3,000 ton wooden freighter."[17]

The officials of the United States Shipping Board, organized in 1916, before the American entrance into the conflict, argued among themselves the merits of wooden ships. Chairman William Denman was a strong supporter, but Major General George W. Goethals (of Panama Canal fame) thought that all priorities should go to steel ships.[18] As it turned out, all kinds of ships were needed, and many of both varieties were built as quickly as possible.

With the declaration of war in April, 1917, the Shipping Board redoubled its efforts to resolve the shortage of ships. The Emergency Fleet Corporation (EFC), which was to own and operate the ships that were to be built, called for construction of a wooden merchant fleet that would achieve a rate of 200,000 tons a month by October, 1917. It envisioned ships of 3,000 to 3,500 tons each, powered by oil or steam engines capable of making twelve knots. Eight hundred to one thousand of these ships were to be completed within the next fourteen to sixteen months.[19] To stress the need for haste, the editor of the *Gulf Coast Lumberman* called for dedicated efforts by shipbuilders and workers alike:

The building of wooden ships in huge numbers by the emergency route as a means of off-setting the submarine activity of the Germans, occupies the

---

[16] *Gulf Coast Lumberman*, January 1, 1917, p. 26.

[17] Winthrop L. Marvin, "A Thousand Wooden Ships for War Trade," *American Review of Reviews* 55 (May, 1917): 519–20.

[18] Arthur S. Link, *American Epoch*, p. 205; Edward N. Hurley, *The Bridge to France*, pp. 27–28.

[19] Marvin, "A Thousand Wooden Ships for War Trade," p. 520.

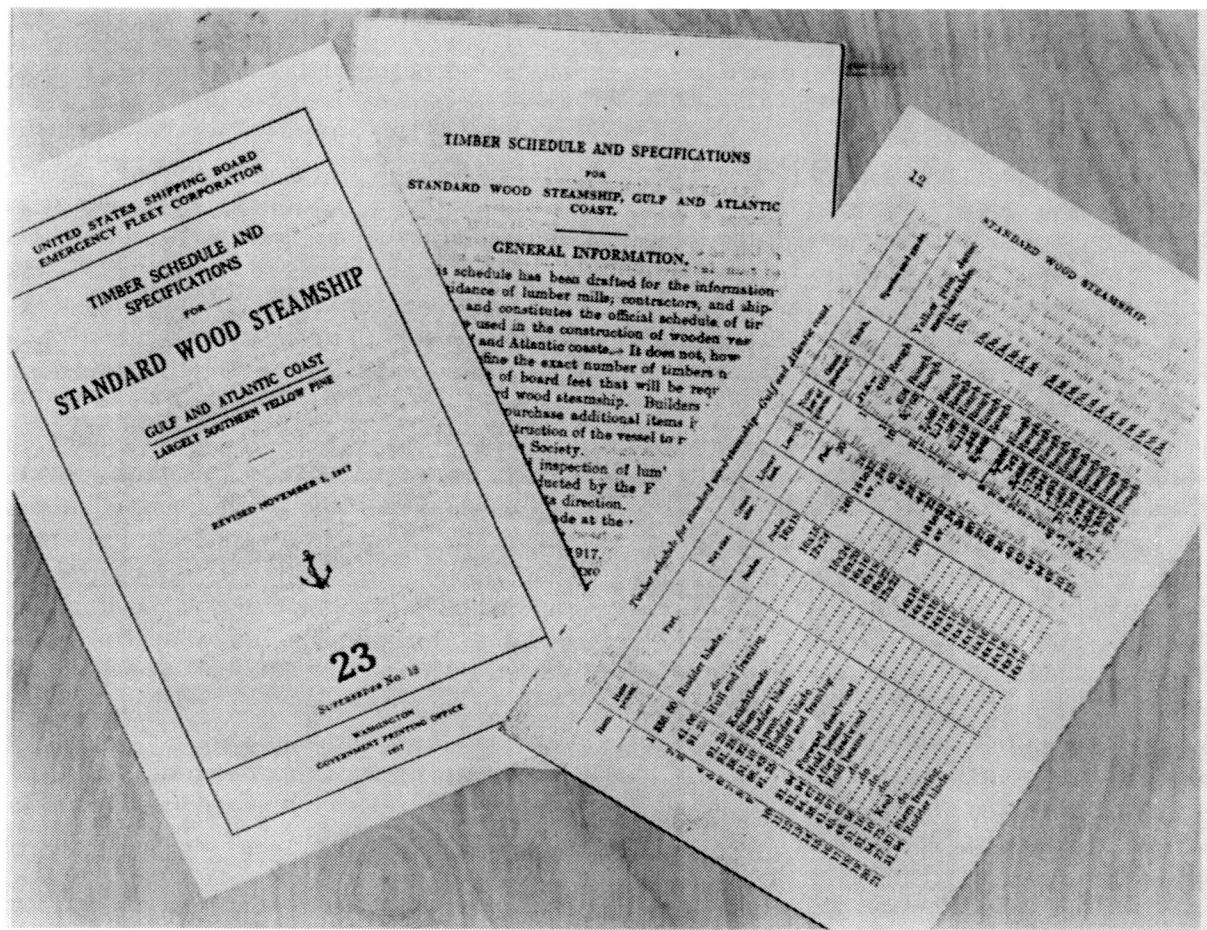

U.S. Shipping Board specifications for the "Standard Wood Steamship," 1917. *Courtesy of Stephen F. Austin State University, Forest History Collections.*

center of the lumber stage at the present time and will probably do so for some time to come.

Practical recognition of the fact that the submarine menace must be counteracted if the nation is to prosper, and the energetic action taken by the Government to secure the building of ships on a tremendous scale, has had its effect. The brains of the nation are today being directed to the business of discovering ways and means for turning out ships, ships, and still more ships.

Since speed in ship construction can be obtained in wooden construction more than in steel, wooden ships by the hundreds and thousands are being called for. Every shipyard is going to be taxed to capacity for the next two years. Every ship carpenter, every man capable of becoming a ship carpenter, is in great demand. There is work for everyone.

New ships must be built and started into action.

The entire Gulf coast, the Atlantic coast, and the great line of the Pacific coast, will be lined with ship yards. The Government has arranged already with the wood manufacturers of the nation, for the securing of plenty of wood for the construction of thousands of ships.[20]

Much of the wooden-ship construction took place on the Gulf Coast, and the three Texas Gulf ports Houston, Beaumont, and Orange made significant contributions. As has been seen, companies in all three ports had been ac-

[20] *Gulf Coast Lumberman*, May 15, 1917, p. 5. To aid lumbermen, contractors, and shipbuilders with the specifications for wooden ships, the Emergency Fleet Corporation published a booklet, *Timber Schedule and Specifications for Standard Wood Steamship*, copy in Forest History Collections, SFA Library.

tively engaged in shipbuilding before the outbreak of the war, and each had installed new equipment and facilities. Now these facilities were to be expanded still further, and government orders were to take priority over all others. To procure the necessary lumber and large timbers for this massive program, the federal government on November 1, 1917, preempted all lumber more than two inches thick, ten inches wide, and twenty feet long that could be used to fill a ship's schedule. The Southern Pine Emergency Bureau, an agency of the Southern Pine Association, cooperated with the order fully and urged its member companies to support the program and to save their scale sheets (records of the size and length of the day's cut) so that government inspectors could see that the directive was being observed.[21]

Soon large wooden ships were being built on the Gulf Coast. On the Houston ship channel the Midland Bridge Company and the Universal Shipyard Company, later joined by Kirby's Houston Shipbuilding and Dry Dock Company, took contracts with the EFC and hastened to begin turning out the standard (the so-called Ferris-type, 3,500-ton) wooden ship. At Orange the National Ship Building Company undertook the construction of vessels even larger than the standard wooden ship. The Orange Maritime Corporation and the Piaggio shipyard accepted contracts from the EFC. In Port Arthur the Long-Bell Lumber Company acquired a site and built a shipyard to construct four or more wooden ships.[22]

At Beaumont the Lone Star Ship Building Company contracted for four ships, and the Beaumont Ship Building and Dry Dock Company undertook to build eight ships for the EFC. This last organization was a most ambitious undertaking. Owned by lumbermen John Henry Kirby, J. W. Link, and B. F. Bonner, and with C. O. Yoakum as general manager, the company occupied a sixty-seven-acre

tract known as Industrial Island. The Beaumont city fathers gave the land, owned by the city, to the Kirby group in December, 1917, "to assure the early establishment of a large ship building industry for Beaumont." The company spent more than $30,000 on the site and about $150,000 on buildings and equipment and within four months had constructed eight marine ways. General manager Yoakum was optimistic concerning the company's prospects:

We now have a present working force of 260 men and expect to more than double this in the next 90 days . . . so that the work on the six hulls now being built may be facilitated. Our payroll is $20,000 per month—by March 1, 1918, it will be $80,000. The plant will expand further and within 90 days two more keels will be laid; thus all eight ways will be in use.[23]

Initial confusion and misunderstandings were inevitable. A notice from the EFC that it would allot ship contracts on a lump-sum instead of a fee (building costs plus a fee for labor) basis brought a temporary hestitation by the builders until the order was clarified and modified.

Many companies had difficulty supplying the large-size timbers required. In July, 1917, Major General Goethals threatened "summary action" because all but twenty-three mills of the eighty-four that had been assigned lumber schedules for shipbuilding had rejected the schedules. It was speculated that perhaps the government would commandeer the mills. The Southern Pine Emergency Bureau worked out a solution by which the orders were distributed among the mills on a more acceptable basis, and the crisis was averted.[24]

To expedite further the flow of lumber and timber products to shipyards, Edward N. Hurley, chairman of the United States Shipping Board, appointed John Henry Kirby to the post of lumber administrator for the South. It

[21]*Gulf Coast Lumberman*, November 15, 1917, p. 22.
[22]*American Lumberman*, May 5, 1917, p. 56; May 19, 1917, p. 34; May 26, 1917, p. 32; July 14, 1917, pp. 36–37.

[23]See note 22 above. *Beaumont Enterprise*, December 5, 16, 1917.
[24]*American Lumberman*, June 9, 1917, p. 30; July 28, 1917, pp. 33–34.

was Kirby's task to secure the cooperation of the millowners with the board in lumber production for shipbuilding. Kirby, as a past president of the National Lumber Manufacturers Association, was a logical choice and was in a position to wield a powerful influence in support of the Shipping Board. He served with considerable success and not a little controversy from March to late July, 1918, supervising the purchase of lumber supplies in the South and facilitating production and delivery of ships' timbers to the yards in accordance with the board's schedule.[25]

During the war years the lumber industry, both in Texas and throughout the South, enjoyed an unprecedented boom. With the wartime orders prices rose from $12 per thousand board feet (for the average mill run) in 1915 to $14 in 1916. With the entrance of the United States into the war in April, 1917, and the government's demand for quantities of lumber of all types, prices spiraled upward in spectacular fashion.[26]

Lumber spokesmen could not conceal their satisfaction. Under the heading "Lumber Prices Break All Records," one newspaper editor reported that the most common remark made by lumbermen during the past few weeks had been, "I never saw anything like these lumber prices." "As a matter of fact," continued the editor, "there is no price on lumber, but cars loaded with lumber are strictly at a premium. . . . Yellow pine lumber in the southwestern territory today is worth almost anything that the man who can make delivery is

minded to ask for it."[27] The average price of yellow-pine lumber, he estimated, was between $25.00 and $26.00 per thousand board feet at the mill, and B&B flooring was bringing $32.50 at the mill. A year earlier, he reminded his readers, the price had been about $20.00. Continuing the comparison, the editor quoted additional prices:

2×4-16 is worth about $24 or $25 right now, as against $15 at this time last year, and 2×10-12 are bringing about $22.50 as against $14 or $15 last year. 4×4 and 4×6 stuff is worth seven or eight dollars more than a year ago. Big timbers are high as a camel's ears, and special cutting is worth a lot of money, depending a good deal on the service and delivery required. Edge grain flooring is getting in the oak flooring class, and battleship decking prices sound like fairy stories.[28]

Prices were indeed favorable. In 1918 the federal government, over the protest of southern lumbermen, fixed the price of southern pine, as well as of Douglas fir and other timbers on the West Coast. For example, the price for yellow-pine boards, 1 by 12 (s2s [dressed on two sides]) B&B, was $31.50. As on wheat, corn, and other commodities, federal war agencies established controls on lumber products and to a large extent stabilized the market.[29]

A shortage of railroad freight cars further complicated the picture. The lack of adequate cars, a regular complaint of lumbermen since the 1890s, was aggravated by the sudden and unprecedented demands for cars by all war-oriented industries. As one spokesman complained, the lumber was cut and ready, but it was no longer a question of price. It was "merely a question of whether the material can be delivered and in most instances it can not. . . . Our company ordered eight cars per day —we received two. However, the government orders are still moving promptly."[30]

[25] Ibid., March 16, 1918, p. 52. Much of the controversy centered about Kirby's protest against the fixing of prices for southern pine lumber by the War Industries Board. Soon afterward the Shipping Board reorganized and eliminated the post of lumber administrator. Kirby at once resigned. J. L. Ackerson to J. H. Kirby, July 16, 1918, Kirby Papers, University of Houston Library, Houston; Gulf Coast Lumberman, August 1, 1918, p. 6; August 15, 1918, p. 9. See also James E. Fickle, The New South and the "New Competition": Trade Association Development in the Southern Pine Industry. Fickle concluded (pp. 93–105) that Kirby was, in effect, fired.

[26] Gulf Coast Lumberman, August 15, 1914; American Lumberman, November 11, 1916, p. 56; May 5, 1914, p. 65.

[27] Gulf Coast Lumberman, May 15, 1917, p. 5.

[28] Ibid.

[29] American Lumberman, April 6, 1918, p. 46.

[30] Beaumont Enterprise, December 16, 1917, p. 12.

THE NEW RECRUIT

Cartoon depiction of John Henry Kirby's war effort. *Courtesy of* Gulf Coast Lumberman.

Labor also profited from the boom. As might be expected, the wages of skilled operators rose more than those of unskilled laborers. According to the standard authority on East Texas labor, wages paid to sawyers in Texas increased from $.55 an hour in 1915 to $.83 in 1919. In a sixty-hour week this represented a raise in total pay from $33.00 to $49.62. The pay of the average unskilled worker also increased, by perhaps one-third. In 1919 some mills reported weekly pay to the "laborer group" at $15.62 a week, as compared to $9.41 in 1915. Day laborers' wages went up from an average of about $1.75 to approximately $2.50 a day.[31]

Because of the proximity of the Gulf Coast shipyards, many East Texas sawmills suffered from a shortage of labor. Some workers quit their sawmill jobs to get better-paying positions in Houston, Beaumont, and Orange. Advertisements in East Texas and metropolitan papers emphasized the shortage of skilled labor.[32] At Kirbyville, and no doubt elsewhere, millowners put up posters appealing to the patriotism of their workers to stay on the job:

To Lumbermen: For the support of our soldiers in France, the government must have wooden ships. Without ships the war cannot be won. Without timbers ships cannot be built.

OUR COUNTRY LOOKS TO YOU!

Every swing of an ax, every cut of a saw, may score as heavily as a shot fired from the trenches.

Help our boys in France.

Help them to win the war.

MAKE THE WORLD SAFE FOR DEMOCRACY.[33]

The shipyard workers were far more militant than the workers in the sawmills and logging camps. About sixteen hundred men were employed at the shipyards in Beaumont alone during the peak of the wooden-shipbuilding program, and this proximity prompted the ex-change of grievances and determination to take action in the face of a rising cost of living and a shortage of adequate housing. There were strikes and threats of more strikes, but adjustments of wage scales and mediation by representatives of the United States Department of Labor averted any prolonged tie-up of the shipyards and ensured that the wooden-ship program proceeded on schedule.[34]

By the end of 1917 the first of the wooden ships were almost ready to come off the ways. In Houston, Beaumont, and Orange the rival firms raced to be the first to launch a ship under the Shipping Board's program. From press accounts it appears that no ships (other than tugs and tenders) were completed before the end of 1917 but that all the major shipyards launched one or more vessels during the early months of the next year.[35]

In Orange, where ships had been built since the Civil War, the shipyards were an imposing sight. A reporter recorded his impressions so vividly that one can almost feel his excitement:

The National Ship Building Company is engaged in building fourteen wooden ships for the United States Government, and it was the size and proportions of these that opened the eyes of the lumber editor. Far larger in every way than the big standard wooden ships that the Government is building elsewhere, the boats on the ways at this yard fairly startle the beholder.[36]

Two of the ships, continued the reporter, were practically closed in, and one was almost ready to launch. The observer climbed a stair up the side of the ship, but it was not until he reached the top and looked down into the interior that the great size of the vessel became apparent:

To the uninitiated, the tremendous construction of these big vessels is a great surprise. You look down into the great hold, and see a scientific arrangement

[31] Ruth A. Allen, *East Texas Lumber Workers: An Economic and Social Picture, 1870–1950*, pp. 89–92.

[32] *Beaumont Enterprise*, December 1, 1917.

[33] *Beaumont Enterprise*, December 16, 1917, p. 11.

[34] *Beaumont Enterprise*, December 1, 1917, p. 7; December 5, 1917, p. 10. *Houston Chronicle*, January 11, 1918, p. 9; January 13, 1918, p. 1; *Gulf Coast Lumberman*, November 15, 1917, p. 32.

[35] *Houston Chronicle*, January 5, 1918, p. 1; January 18, 1918, p. 9.

[36] *Gulf Coast Lumberman*, November 15, 1917, p. 30.

The *War Mystery. Courtesy of* Gulf Coast Lumberman.

First standard wooden ship built on the Gulf Coast, the *Nacogdoches*, is launched on the Houston Ship Channel. *Courtesy of* Gulf Coast Lumberman.

of huge timbers that take your breath away. The keel is perfectly huge, and gives the visitor a feeling of strength and stability inexplainable. For every piece of timber or lumber is dipped or painted with creosote before using, so that the hull is black and imposing.[37]

According to officials at the National Ship Building Company, the keel of this ship was laid in July, 1917, and they expected to launch it by December 15. Thereafter they planned to build ships of this size in ninety days. Since eight to fourteen ships would be built at the same time, this would mean a ship launching every few weeks at this one shipyard.[38]

The National finally launched its "great ship" on February 27, 1918. Described as the "largest wooden ship that ever took water," the vessel was 330 feet long with a 48-foot beam and a 27-foot molded depth and carried 5,000 tons. Christened *War Mystery*, the ship actually made her descent to the Sabine several minutes before schedule. A reporter described the event:

Suddenly she started into life. The whistle of the shipyard blew a shrill warning, and the nine hundred men who were helping in the launching prepa-

rations jumped clear of trouble. Smooth as glass, without a hitch or trouble, the great boat slid down her ways. As she struck the water, heading at great speed for the opposite side of the river, four great anchors that were held on trigger attachments, were turned loose, and her progress was so carefully checked that she was within forty feet of the opposite bank before she came to a dead stop. Then she swung upstream with the strong wind that was blowing, and floated beautifully—a masterpiece.[39]

The National worked closely with Lutcher and Moore Lumber Company, which had the contract to supply the shipbuilder with most of its timber and lumber materials. Situated between the two L&M mills on the west bank of the Sabine, it was a natural and profitable arrangement for both companies. A. A. Daugherty, of National, had only the highest praise for the L&M operation, which provided him timbers "on time and in the right order." In turn, officials of L&M praised Daugherty as a "high-class engineer" who brought efficiency to the shipbuilding business and made the National yard in Orange "the leading wooden ship building point in the United States."[40] To-

---

[37] Ibid.
[38] Ibid.

[39] *Gulf Coast Lumberman*, March 1, 1918, p. 32; *American Lumberman*, March 2, 1918.
[40] *Gulf Coast Lumberman*, November 15, 1917, pp. 30–32; March 1, 1918, p. 32.

gether they pledged to build the "bridge of boats" needed to win the war.

The National was not the only shipbuilding yard in Orange. The Southern Dry Dock Company contracted with the EFC to build five standard-type wooden steamships. Across the river in East Orange, Louisiana, the International Shipbuilding Company had contracts with the EFC for a number of wooden ships, and the Orange Maritime Corporation was also building vessels of the same type. According to one reporter, the "whole waterfront at Orange was a cluster of shipyards." In November, 1917, he counted thirty-five ships being rushed to completion in Orange alone. He concluded that Orange was certainly doing its best to "win the war with ships."[41]

In Houston the shipyards also began launching vessels early in 1918. The Universal Shipbuilding Company, on the Houston ship channel, built standard wooden ships for the EFC under the general supervision of Charles N. Crowell, district superintendent for the western Gulf Coast. Early in April, Universal announced the launching of the *Nacogdoches*, which was the occasion for a gala celebration. On the first Saturday in April, Nina Cullinan, the daughter of J. S. Cullinan, a Houston oil magnate and civic leader, christened the ship just before it slid down the ways into the channel. The name *Nacogdoches*, which had been chosen by Mrs. Woodrow Wilson, was particularly appropriate for a vessel built of yellow pine from the East Texas Piney Woods.[42] During the same month George Cole, of the Midland Bridge Company, launched the first of six wooden ships that he had contracted to build for the EFC. It was christened the *Houston*, and its launching was accompanied by a suitable ceremony and celebration.[43]

Shipbuilding activities in Beaumont fol-

lowed the pattern set in Orange and Houston. No less than six Beaumont shipbuilding companies were busily turning out standard wooden ships for the EFC. Early in May, 1918, the Beaumont Ship Building Dry Dock Company launched the *Oneco*, the first ship built on the Neches.[44] Later in the same month the company launched the *Swampscott*, and the Lone Star Ship Building Company launched the *Lone Star* on June 1. The McBride and Law Company, one of the larger shipbuilders on the Neches, launched the *Beaumont* early in July and undertook to meet a schedule by which it would launch one ship a month thereafter.[45]

In addition the Lone Star Ship Building Company received a contract from the EFC to equip for sea duty twenty-two wooden ships built in the Beaumont area and elsewhere. The Beaumont Ship Building Company contracted to equip thirty ships. Thus the two Beaumont-based firms would provide equipment and prepare for sea duty more than fifty ships built and launched on the Gulf Coast.[46]

The armistice of November 11, 1918, brought the war to a sudden end and caught shipbuilders at the peak of their planned production. Fourteen ships were on the ways in Houston, sixteen under contract in Orange, and about twenty in various stages of construction in Beaumont. The Emergency Fleet Corporation stopped issuing new contracts and canceled others. Yet the shipyards built, launched, and delivered more ships in 1919 than they had in 1917 and 1918. For example, the Beaumont shipbuilders did not launch the Ferris-type wooden ship *Angelina* until March, 1919; it sailed on its first voyage in August of the same year.[47]

The overall record of performance by the

[41] *American Lumberman*, March 2, 1918, p. 39.
[42] *Nacogdoches Daily Sentinel*, April 3, 1918.
[43] *Houston Chronicle*, January 13, 1918; *Nacogdoches Daily Sentinel*, April 3, 1918. According to Superintendent Crowell, seventeen ships had been launched before April 1 in his district, which included Louisiana and Texas.

[44] *Gulf Coast Lumberman*, March 1, 1918, p. 32; *American Lumberman*, August 3, 1918.
[45] *American Lumberman*, July 6, 1918, p. 46; August 3, 1918, p. 35.
[46] *American Lumberman*, March 2, 1918, p. 57.
[47] *American Lumberman*, November 30, 1918, p. 73; December 7, 1918, p. 54; Paul M. Zies, *American Shipping Policy*, pp. 95–114.

wooden-ship division of the Emergency Fleet Corporation is mixed and subject to interpretation. No wooden ships were delivered to the Shipping Board in 1917. At war's end 116 wooden ships had been delivered, 86 of which were in operation and were carrying cargo. The next year, 1919, was the peak year for the EFC. The wooden-ship division reported that 403 ships weighing more than 1,338,000 deadweight tons had been built and delivered to the government. The accounts of "Building the Emergency Fleet" broke down construction records not by individual yards but by the four districts in the wooden-ship division: the Pacific, Atlantic, Southern, and Gulf (west of New Orleans). The last included the yards in Texas and a few in western Louisiana. If, as appears reasonable, the Texas yards built approximately one-fourth of all the wooden ships, the output of the Texas yards exceeded 20 ships in 1918 and more than 100 ships in 1919. If the war had lasted through 1919, as most officials expected, that would have been a significant contribution, a total of more than 440,000 deadweight tons of new shipping. The total wooden-ship production in the United States for the war-emergency period was 589 ships weighing 1.885 million tons. Although modest beside the more than 7.7 million deadweight tons (1,200 ships) produced by the steel-ship division, the wooden ships would have provided valuable additional tonnage that might have proved crucial in the battle to span the Atlantic with a "bridge of boats."[48] Viewed in this light, the efforts of the pine lumbermen of Texas and the Gulf Coast shipbuilders represented no mean achievement in the war effort to "make the world safe for democracy."[49]

[48] W. C. Mattox, *Building the Emergency Fleet*, pp. 47–48, 104–105; Edward N. Hurley, *The New Merchant Marine*, p. 51. "Report of the United States Shipping Board for 1919," in *Literary Digest*, March 27, 1920, p. 153; U.S. Shipping Board, *Fourth Annual Report of the United States Shipping Board*, pp. 204–205.

[49] In addition to its shipbuilding efforts, the southern-pine industry, including that in Texas, provided great quantities of lumber products for all other phases of the war program. According to the Southern Pine Association the industry delivered almost 2 billion feet of lumber to the United States and the Allied governments. See Fickle, *The New South and the "New Competition*," p. 104.

CHAPTER 13

"Cut Out and Get Out"

There is no sound so sad as a silent mill
whistle.—Dr. George F. Middlebrook

THE postwar boom continued into 1920, and Texas sawmill owners shared in the general prosperity. Prices for mill-run lumber ranged as high as $42 per thousand board feet, which was higher than the peak prices of the war years. Then the market dropped dramatically and became stabilized near $30 a thousand, where it remained for the next several years. Timber resources declined rapidly, and by the end of the decade forestry experts were estimating that fewer than 1 million acres remained of the great virgin forest that once covered 14 to 18 million acres of East Texas. Builders turned increasingly to the West Coast for quality-lumber needs as experts predicted that Texas would soon be transformed from a major lumber-exporting state into a lumber-importing state. Few new mills were built in Texas, and many of those that burned were not rebuilt. Most large manufacturers began contemplating the course of action they would take when their timber supply was exhausted.[1]

What was perhaps the last new large mill in East Texas was constructed in 1917.[2] It was an interesting and unusual project involving the Lutcher and Moore heirs and Robert W. Wier, an experienced Houston lumberman. The Lutcher heirs—Mrs. Henry Lutcher, Miriam Lutcher Stark, William H. Stark, Carrie Lutcher Brown, and Dr. Edgar W. Brown—held an area of about 86,000 acres of virgin longleaf in northern Newton, Jasper, and Sabine counties. The timber lay seventy-five to eighty miles north of Orange, and owners considered it uneconomical to cut and transport logs to their own mills. They contracted with Wier to build a sawmill on the site and cut the timber there. Wier was to build a mill capable of cutting 40 million board feet a year and to pay the Lutcher heirs a stumpage fee of $6.00 per thousand board feet for all timber cut plus 25 percent of all sales at prices over $13.50 per thousand board feet. It was expressly provided that Wier was to clear-cut the acreage of all merchantable timber over 10 inches at the butt. Further arrangements were made for conveying to Wier title to lands occupied by the millsite and provisions for the building of tramways and railroads.[3]

Wier built the mill and fulfilled the contract. His mill was a complete, modern outfit designed to cut 200,000 board feet in a ten-hour

[1] Conversation with W. S. Brame, Livingston, March, 1960, Oral History Collections, Forest History Collections, Stephen F. Austin State University Library, Nacogdoches; hereafter cited as SFA Library. Unless otherwise noted, interviews and conversations cited are in these collections, and all interviews were conducted by principal author Maxwell. Statement of Operating Costs, 1924–39, Peavy-Moore Lumber Company, Deweyville, Miscellaneous Papers, Forest History Collections, SFA Library.

[2] To the authors' knowledge, this appears to have been literally true. Excluding replacements of burned or worn-out mills and consolidations (such as, in 1955, the Kirby mill at Silsbee, which replaced several other Kirby mills), no new large mills

(that is, mills with a capacity of more than 100,000 board feet a day) have been built in Texas since the construction of the Wier mill in 1917.

[3] Contract between R. W. Wier with Stark and Brown Interests, February 12, 1918, Lutcher and Moore Lumber Company Papers, Forest History Collections, SFA Library.

day. The mill consisted of two double-cutting-band rigs and a circular saw. He also built a company town, Wiergate, for his workers, sheds, dry kilns, a planing mill, a commissary, and a short-line railroad, the Gulf and Northern Railroad, which ran seventeen miles south to Newton, where there was a junction with the Orange and Northwestern Railway. In this remote latter-day barony the Wier Longleaf Lumber Company logged what was perhaps the last great Texas stand of longleaf pine. The company used a tramroad and spur lines with steam skidders and loaders to log the area in near-record time. In accordance with the contract the company clear-cut the area, leaving only small pines, a growing jungle of scrub hardwood saplings, and a great mass of discarded limbs and tops.[4]

Many lumbermen believed that the cutover lands could profitably be used for general farming. If this was feasible, then it was desirable to clear-cut the entire block and leave no trees to impede the plow, and the use of the skidder could be justified. Perhaps the most confident of the large operators was Robert A. Long, of Long-Bell Lumber Company, who held large areas of cutover land in both Texas and Louisiana. Boldly proclaiming that "the plow follows the skidder," he organized the Long-Bell Farm Land Corporation to sell small farms of 40 to 80 acres each to northern and western farmers and stockmen. To publicize the project, Long set up a model farm in western Louisiana near De Ridder, where he demonstrated that fruits, vegetables, and row crops could be raised. Long argued that his experiments in Wisconsin, where he had converted cutover lands near Antigo into profitable farms, proved that Texas pinelands would also be good for farming. The company, which operated sawmills in Doucette and Lufkin, as

well as several mills in Louisiana, sold about 300,000 acres of land to several thousand farmers from northern states and the West, and to recently arrived immigrants from Europe.[5]

Robert Long was not the only entrepreneur who sought to transform cutover pinelands into fertile farms. In 1914 a Chicago syndicate purchased about 12,000 acres of land in Liberty County, planning to colonize it with Greek families who would raise grapes, fruit, and truck-garden products for the Houston and Galveston markets. The company established a town, appropriately called Macedonia. The outbreak of World War I prevented this venture from materializing.[6] The Foster Lumber Company sold a number of small farm sites to immigrants from central Europe, and William Carlisle organized the Onalaska Livestock Company to promote the sale of cutover lands for ranches. The Santa Fe Railroad sold farm sites from cutover acreage along its East Texas branch line, and the Kirby-dominated Southwestern Land and Development Company offered cutover lands for sale to farmers, ranchers, and fruit growers.[7]

Almost without exception these promotions were failures. J. W. Pedigo, of the Southwestern Land and Development Company, remarked that his company had very little success selling cutover lands for farm plots. Most families that did purchase farms abandoned them within a few years and sought employment in Houston, Galveston, or Beaumont industries or in the oil fields. The light, sandy soil of the East Texas Piney Woods was suited for growing trees but very little else. Most of the acreage involved in these farmsite schemes has long since reverted to forest.[8]

---

[4] *American Lumberman*, May 3, 1919, p. 8; conversation with A. E. Cudlipp, Lufkin, March 8, 1963; conversation with C. J. Smith, Wiergate, July 30, 1963. Wier cut out and suspended operations in 1943, twenty-five years after he began fulfilling his cutting contract. The mill and cutover acreage were sold, and the large mill was dismantled. A modest-sized mill continued to operate on the second-growth forests.

[5] *Log of Long Bell*, January, 1919; February, 1919.

[6] St. Clair G. Reed, *A History of the Texas Railroads and of Transportation Conditions under Spain and Mexico and the Republic and State*, p. 480.

[7] Foster Lumber Company Papers, Forest History Collections, SFA Library; *American Lumberman*, December 16, 1916, pp. 40–41; conversation with J. W. Pedigo, Jasper, August 12, 1963.

[8] William T. Chambers, "Divisions of the Pine Forest Belt of East Texas," *Economic Geography* 6 (January, 1930): 94–103.

Cutover scene, after logging and burning. *Courtesy of U.S. Forest Service.*

In the decade after World War I a large number of companies cut out and suspended operations. Many of them were large, well-established, widely known mills that employed hundreds of workers and whose annual cut ran into the tens of millions of board feet. Among the first of the big mills to discontinue operations was the Central Coal and Coke Company (the 4-C), which shut down soon after the end of the war. It sold its properties to the Houston County Timber Company, which dismantled the large mill and replaced it with a much smaller rig. It continued to cut the remaining timber and in turn closed down with the coming of the Great Depression.[9]

In reminiscing about the bonanza days in the lumber industry, Al Brown, formerly an editor with the *Gulf Coast Lumberman*, recalled that "along about 1924 a significant change began to take place in the lumber industry. More and more of the larger sawmills

shut down. Out of timber." This trend not only continued but accelerated. In January, 1927, the *Gulf Coast Lumberman* published a list of closed-down mills whose total capacity had run to the staggering figure of 3.95 million board feet a day. In addition to the 4-C, the list included the Thompson-Ford Lumber Company, the South Texas Lumber Company (with three mills), the W. M. Cady Lumber Company, the W. G. Ragley Lumber Company, the Alexander Gilmer Lumber Company, Lock-Moore and Company, the Trinity River Sawmill Company, the Long-Bell Lumber Company (four mills), the Industrial Lumber Company, and the Enterprise Lumber Company.[10]

In 1925 the big Onalaska mill of William Carlisle and James West in Polk County cut out, and its owners moved their operations to the West Coast. Onalaska became little more than a ghost town. The last of the Texas-based

[9] Conversation with H. Z. Collier, Kennard, July 29, 1964; conversation with Mary Aldrich, Crockett, July 29, 1964.

[10] Al Brown, "This I Remember," *Gulf Coast Lumberman*, November, 1963, p. 30.

Cutover acreage, logging operations not yet complete, Trinity County. *Courtesy of Stephen F. Austin State University, Forest History Collections.*

Long-Bell mills, the Lufkin Land and Lumber Company, abandoned operations in 1930, and that giant corporation also transferred its lumbering activities, except for its retail yards, to the West Coast.[11]

In the same year, 1930, the great Trinity County Lumber Company discontinued operations. The owners, the Joyce family, had been in Texas about forty years and claimed to operate the largest and oldest sawmill in the South. The mill had a capacity of 300,000 board feet of lumber in a two-shift, twenty-four hour day and employed 200 to 400 men. As could be expected, it dominated the economy of Groveton and much of Trinity County. The

founder of the company, David Joyce, had developed both sawmills and retail yards in the Middle West, with headquarters in Iowa. His son, William T. Joyce, expanded those properties and in 1890 moved to Texas, where he purchased both a sawmill and more than 75,000 acres of virgin timber and constructed a railroad (the Groveton, Lufkin, and Northern) to give him an alternate outlet for his products. William's son, James Stanley Joyce, graduated from Yale and then assumed active management of the Texas properties (he was perhaps better known, however, as the one-time husband of Peggy Hopkins Joyce, a Broadway and motion-picture actress; their much-publicized marriage lasted only about a year). The Joyce mill in Texas cut an estimated total of 2 billion board feet, using tramroads, spur lines, and rehaul skidders that denuded the land and left the company's holdings a tangled mass of

[11] Conversation with W. S. Brame, Livingston, 1960; K. G. Hanson to W. D. Oliver, Kansas City, June 5, 1958, Miscellaneous Papers, Forest History Collections, SFA Library. See also John M. McClelland, Jr., *R. A. Long's Planned City: The Story of Longview.*

stumps, treetops, and scrub undergrowth. The shutting down of the Trinity County mill brought instant depression to the town and the surrounding county. The town of Groveton declined, and the railroad was abandoned and its rails taken up. The workers moved away to seek other employment or tried to scratch a living as small farmers. Stanley Joyce simply returned to the family home in Chicago, where he managed his other interests.[12]

In 1931 the Pickering Lumber Company cut out and shut down operation of its large mill at Haslam, Texas. Natives of Missouri, the Pickerings developed a large lumber business including manufacturing and wholesale and retail branches, with headquarters in Kansas City. The company had properties in Missouri, Arkansas, Oklahoma, Louisiana, and Texas. In addition to the Haslam mill, the Pickerings owned more than 100,000 acres of timberland in Shelby, Sabine, and San Augustine counties. From the time the Haslam mill was built in 1913, the owners exhibited great haste in converting the timber resources into lumber. They ran a two-shift, twenty-four-hour operation, cutting about 200,000 board feet a day. Logging was done almost entirely by tramroad with spur lines extending into the forest at intervals and steam-powered rehaul skidders delivering the logs to trackside. Veteran employees agreed that the Pickering Company did not practice conservation or careful management. The logging methods destroyed most of the younger trees and seedlings. The cutover areas resembled a vast wasteland. The founder, W. R. Pickering, apparently seldom if ever visited the Texas holdings, and his son, W. A. Pickering, inspected the operations only infrequently. Like other absentee owners, they left

management details to their salaried employees and were chiefly concerned with production figures and earnings. The Pickering company had previously ceased operations at three mills east of the Sabine in Louisiana, and the closing of the Haslam mill terminated their interest in the Gulf Southwest. The Pickerings transferred their activities to the Pacific Coast, where they engaged in large-scale lumbering.[13]

What happened to a mill town when the owners cut out and discontinued operations? According to veteran mill employees, the first reactions were shock and disbelief. Usually there had been rumors that the owners were planning to close the mill, but the rumors were usually followed by denials from the local officials and reassuring statements that the mill would continue. Then suddenly, often on a Monday morning, it was all over. The mill whistle did not blow, and no logs went up the jack ladder and through the saws. Men and women stood in the company streets talking and trying to make wry jokes about the situation. Wives and mothers were often tearful at the realization that the family was at loose ends and the seemingly all-powerful and paternalistic company had abandoned them. Workers who had spent several years with the company, put down roots in the town, and risen above the status of laborer often found it hard to find other employment. By the late 1920s the general decline in sawmilling in Texas had created a surplus of sawyers, edgermen, trimmers, loader foremen, woods bosses, and other specialists. Companies that continued operations, such as Temple, Angelina County, Kirby, and Carter, often noted that they had several experienced sawyers working in lesser capacities, sometimes as common laborers. The 25 to 30 percent of the work force that was described as "always going or coming" simply moved on to other jobs or to the cities. East Texas is sprinkled with ghost towns and semi-

---

[12] More accurately, both William and Stanley Joyce should be considered absentee owners, for they maintained homes and headquarters in the Middle West. Stanley Joyce, however, also had an apartment in Groveton and spent much time there. See Adele Mansell, "A History of Trinity County" (master's thesis, Sam Houston State Teachers College, Huntsville, 1941), pp. 60–61; James E. Defebaugh, *American Lumbermen*, 3 : 165–68; conversation with Roy Dudley and W. L. Avery, Groveton, March 15, 1960.

[13] *American Lumberman*, April 21, 1906; June 16, 1917; *Champion* (Center), August 7, 1952; conversation with W. I. Davis, Center, March 20, 1956; conversation with Dr. George F. Middlebrook, Caro, October 15, 1959.

ghost towns that were once thriving sawmill communities. As Dr. George F. Middlebrook, a company medical doctor who had witnessed the end of three major sawmill operations, remarked, "There is no sound on earth so sad as a silent mill whistle."[14]

Perhaps a symbol of the decline of lumbering in Texas was the demise of the "Orphan Katy" branch of the Missouri-Kansas-Texas Railroad. This sixty-six-mile short line, extending from Trinity on the International and Great Northern to Colmesneil on the Texas and New Orleans, had been constructed in 1882 to serve the sawmills that had sprung up in this fine longleaf-pine belt. So thick were the mills along the route that the engineers said that they were never out of the sound of a sawmill whistle. Along the right-of-way at various times stood mills belonging to such prominent leaders in the industry as William Cameron, Stanley Joyce, John Martin Thompson and his sons, W. T. Carter, W. F. Edens, and W. H. Norris. In the early days it had been a busy, flourishing road with daily freight and passenger trains. By 1930, however, most of the timber had been cut out, and the mills were silent. In 1936 the parent company, now the Waco, Beaumont, Trinity, and Sabine, abandoned the branch line and took up the tracks. Forty years later few people recalled that this east–west line had run through the heart of the longleaf belt or that the town of Corrigan had once been a railroad junction.[15]

The coming of the Great Depression brought an end to the bonanza period of Texas lumbering. Many owners were faced with declining lumber prices and exhausted resources. Even mills that still had reserves of timber reduced output and days of operation to the point that it was practically impossible for workers from

Cartoonist's view on conservation. *Courtesy of Stephen F. Austin State University, Forest History Collections.*

shut-down mills to find employment elsewhere in the industry. From an average production of more than 1.35 billion board feet a year during the first half of the 1920s the annual cut sagged to 1.1 billion board feet in 1929 and less than 900 million board feet in 1930. Not since the 1880s had the annual cut of yellow-pine timber been so small.[16] As the conservationists had warned, Texas had shifted, from a lumber-exporting state to a lumber-importing state. Only a few small, scattered stands of virgin pine remained, and much of the great East Texas pine forest had been reduced to cutover wastes, thickets of scrub hardwoods, and immature second-growth shortleaf, loblolly, and slash pine. It was obvious that the end of an era had been reached in Texas lumbering. The industry had turned full circle.

[14] Conversation with Simon W. Henderson, Jr., Lufkin, August 10, 1964; conversation with Dr. George F. Middlebrook, Caro, October 15, 1959. See also Bob Bowman, *This Was East Texas: An Anthology of Ghost Towns.*

[15] Reed, A History of Texas Railroads, pp. 483–86; *Gulf Coast Lumberman*, November, 1963, p. 21; Robert S. Maxwell, *Whistle in the Piney Woods: Paul Bremond and the Houston, East and West Texas Railway,* pp. 56–57.

[16] Southern Pine Association, *Lumber Production in the South.*

The lumber industry in Texas hit bottom in 1932, coinciding with the depths of the Great Depression. In that year Texas mills produced only 354 million board feet of finished lumber, comparable to production in 1880. The Texas Forest Service estimated that there remained less than 1 million acres of virgin pine timber (much of it in small, scattered tracts) in comparison to approximately 3 million acres in 1916, and 12 to 14 million acres in 1900.[17]

Another reason for the sudden decline in Texas lumber production was the collapse of prices. The cost of producing lumber had remained essentially stable during the postwar decade, ranging from $23.00 to $25.00 per thousand board feet. With selling prices averaging $28.00 to $30.00 per thousand board feet, a small but satisfactory profit could be made. In 1930, however, the price dropped to $24.00; in 1931, to $18.00; and in 1932, to $14.65. Most owners decided to preserve their diminishing timber stocks until there was a general price recovery rather than pay $10.00 per thousand board feet for the privilege of manufacturing lumber.[18]

Owners reacted to the crisis in various ways. Some increased production and offered their lumber for sale at cutrate prices, thus depressing the market even more. Others, who were nearing the end of their timber supply, simply closed the mills and abandoned the cutover lands. Not only were the company's employees suddenly left without employment and often without food, but the county governments found their tax base greatly reduced as the cutover lands became tax-delinquent with no purchasers in sight.[19]

On the other hand, many companies worked out plans to maintain a reduced production schedule and care for their employees. The Frost-Johnson Lumber Company operated its plants on a curtailed basis, and each man shared in the work, even though each might work only a few days a month. To meet the food problem, company officials cleared ground, and the workers put out vegetable gardens to provide food for immediate use and for the next winter as well. Both the Southern Pine Lumber Company (Temple) and the Angelina County Lumber Company (Kurth) ran their mills on a reduced schedule but gave all the employees some work time each week. W. T. Carter and Brother Lumber Company also went on a shortened work week and in 1932 and much of 1933 did not charge its employees house rent.

The Lutcher and Moore Lumber Company, of Orange, the pioneer big mill in East Texas, discontinued manufacturing operations for a time. For about fifteen years the company managed its more than 250,000 acres of timberland in Texas and Louisiana; negotiated oil, gas, and sulfur leases on its property; and maintained a wholesale and retail lumber business in Orange. In addition, to quote H. J. Lutcher Stark, they "raised alligators." Not until after World War II did Lutcher and Moore resume lumber manufacturing.[20]

John Henry Kirby was even less fortunate. Always a large operator whose business ventures had at times bordered on the speculative, Kirby had depended on the continuing economic growth and general prosperity that the United States had enjoyed since 1900 to bring success to his efforts. He lived like a prince and was generous to a fault. A close associate commented that "he was a soft touch, a man who couldn't say no." It was said that he "gave away millions, just helping people." As has

[17] Ibid.; J. H. Foster, H. B. Krausz, and George W. Johnson, *Forest Resources of Eastern Texas*, Texas A&M College, Department of Forestry Bulletin no. 5 (1917), p. 190; *Texas Almanac, 1958–1959*, p. 190.

[18] See Statement of Operating Costs, 1924–39, Peavy-Moore Lumber Company, Deweyville, Miscellaneous Papers, Forest History Collections, SFA Library.

[19] Henry B. Steer, *Lumber Production in the United States, 1799–1946*, pp. 8, 15–16; Henry Clepper, "Crusade for Conservation," *American Forests*, October, 1957.

[20] Conversation with Clyde J. Woodward, Sr., Nacogdoches, October 3, 1957; conversation with Simon W. Henderson, Jr., Lufkin, 1958; conversation with Needham B. Weatherford, Camden, 1958; *San Augustine Tribune*, December 5, 1935; *Beaumont Enterprise*, March 20, 1949; Lutcher and Moore Lumber Company Papers.

been noted in an earlier chapter, the coming of the Depression and the collapse of the lumber market overwhelmed Kirby. His accumulated obligations forced him into bankruptcy, with liabilities listed at more than $12 million. The Kirby Lumber Company continued operation but came under the control of the Atchison, Topeka and Santa Fe Railway. During the last half-dozen years of his life Kirby was little more than a figurehead in the company that he had founded and built into the largest lumber operation in the Gulf Southwest.[21]

[21] *Houston Post*, July 26, 1964. See also personal file in Kirby Lumber Corporation Papers, Forest History Collections, SFA Library.

Neither the state of Texas nor the county governments were able to do much to help the lumber companies or to succor the families of the unemployed or semiemployed workers. At the end of 1932, like depressed people elsewhere throughout the United States, the sawmill communities, workers and owners alike, looked to the coming of a new administration in Washington, with new programs and a bold new optimistic philosophy to reverse the downward spiral of the Great Depression.

CHAPTER 14

A New Beginning

We want our forests reserved for our children's children.—W. Goodrich Jones

THE election of Franklin D. Roosevelt to the presidency in 1932 brought to the White House a person experienced in the problems of forest management and a dedicated champion of conservation. As a young man Roosevelt had taken steps to improve the woodlands on the family estate at Hyde Park, New York, and as governor of New York he had set up conservation projects to provide work for thousands of men on the unemployment rolls. As president-elect he had glimpsed the opportunity for conservation and reforestation to help solve the problems of the Great Depression. Once in office he embarked on a bold program of relief, recovery, and reform measures known as the "New Deal." These measures dramatically affected virtually every person and institution in the United States. Texas and the Texas forests were deeply affected and permanently altered by the New Deal.

By the end of 1932 most American businesses were in deep depression. The lumber industry was prostrate. The Hoover-appointed United States Timber Conservation Board, of which John Henry Kirby was a member, studied the situation and made a series of recommendations for regulation of timber cutting, mergers, and control of production and modification of the antitrust laws. Lack of legal authority, however, fear of antitrust prosecution, and mutual suspicion between trade associations and individual members prevented any effective measures. As William B. Greeley, the former chief of the United States Forest Ser-

vice, said, "The lumber industry had degenerated into a struggle for survival."[1]

Into this situation Roosevelt stepped with a proposal for industrial revival and self-regulation, the National Industrial Recovery bill (later Act). This bill quickly passed through Congress, and Roosevelt established the National Recovery Administration (NRA) under the direction of General Hugh Johnson. The latter immediately began organizing the major industries of the country through a series of fair-practice codes by which industries would pledge themselves to approved trade practices, higher wages, shorter hours, recognition of collective bargaining, and limitation of production. Among the industries called upon to draft an NRA fair-practice code was the lumber industry.[2]

In the South the Southern Pine Association, to which most of the larger Texas manufacturers belonged, became the chief agency for drafting the code for the southern-pine industry and, once it was adopted and approved, administering and enforcing it. Among the principal provisions of the code were production controls, minimum-price schedules, minimum-wage and maximum-hours standards, collective-bargaining procedures, aboli-

[1] Edgar B. Nixon, ed., *Franklin D. Roosevelt and Conservation*, 2:602–605; *New York Times*, April 30, 1933; James E. Fickle, *The New South and the "New Competition": Trade Association Development in the Southern Pine Industry*, pp. 118–19; William B. Greeley, *Forests and Men*, p. 133.
[2] Arthur M. Schlesinger, Jr., *The Coming of the New Deal*, pp. 103–10.

tion of child labor, safety regulations, and a conservation statement. The last provision, Article X of the code, committed the southern-pine industry to approved forestry practices on individual owners' lands, including protection against forest fires, preservation of young trees during logging operations, restocking of the land after timber cutting, selective cutting, and sustained-yield forest-management programs.

Texas lumbermen played important roles in both the drafting and the administration of the NRA lumber code. One member of the SPA committee that drafted the code of the southern-pine industry was A. J. Peavy, of the Peavy-Moore Lumber Company, which had a mill in Deweyville, Texas. He and Ernest L. Kurth served on the lumber-code authority as southern-pine representatives, and were also members of the board of directors for Texas, along with Eli Wiener and R. W. Wier. Peavy also served on the control committee for the enforcement of the code for the southern-pine industry, and the several administrative committees included such representative Texans as Kurth, Paul Sanderson, Herman Dierks, E. A. Frost, and Arthur Temple.[3]

Although most of the larger Texas lumbermen endorsed the code and pledged their support to the SPA as the chief administrative and enforcement agency, the lumber code did not work well. Several hundred new sawmills entered or reentered the industry soon after the code was announced, and each demanded a production quota. The code authority had attempted to set production controls both for the southern-pine industry as a whole and for each region, and production allotments could be made to newcomers only at the expense of the larger mills. Most applicants received production quotas, but not to their satisfaction. The wages-and-hours regulations also shocked many Texas lumbermen, large and small operators alike. Wages doubled, to $.24 an hour,

Planting pine seedlings. *Courtesy of U.S. Forest Service.*

and the work day was set at eight hours. With only a few field workers the SPA found it almost impossible to police adequately the announced minimum-wage schedules. As one operator wrote in the fall of 1933, "Unless we can coerce the unruly fellows into complying as to wages and hours, the fellows who are trying to live up to the law will become discouraged and also become violators."[4] The problem of production allotments was never satisfactorily solved. Even Kurth protested that his Angelina County Lumber Company was hampered by the limitations of the code. In a letter written in March, 1934, he complained that the company had needed a larger quota than the 1.3 million board feet allocated for March. The company had cut and sold that amount of lumber by March 24, and the mill had been idle for the rest of the month. His pleas apparently

[3] "Memorandum on Lumber Code," October 31, 1933, in Nixon, ed., *Franklin D. Roosevelt and Conservation*, 1:213–16; Fickle, *The New South and the "New Competition,"* pp. 122–29.

[4] "Memorandum Relating to the President's Conference with Lumbermen and Foresters," February 23, 1934, in Nixon, ed., *Franklin D. Roosevelt and Conservation*, 1:250–62; Fickle, *The New South and the "New Competition,"* pp. 130–31.

A plantation of young pines. *Courtesy of U.S. Forest Service.*

were unavailing, for, despite frequent letters, the company's allotments were not increased significantly. Throughout the state class B and C mills (the smaller mills) demanded more differentials of production allotments in their favor, while class A mills thought that there should be none at all.[5]

By January, 1935, there was common agreement that the lumber code was not working in Texas or in the other southern-pine states. Large millowners charged that small sawmillers were ignoring the code, including its pricing and production schedules. Small operators countered that the large mills were trying to run them out of business. All groups joined in criticizing the administrator of the code in Washington for vagueness, rigidity, and inconsistent enforcement. The SPA passed a resolution calling parts of the code unworkable and demanding a thirty-day suspension. The whole issue was decided in May, 1935, when the United States Supreme Court unanimously declared the NRA unconstitutional.[6]

Article X of the code, embodying the conservation, forest-management, and reforestation principles that forestry leaders had been urging for a generation, was now stripped

[5] Ernest L. Kurth to Southern Pine Association, March 26, 1934; L. O. Crosby to Charles Green, December 5, 1934; Southern Pine Association to Angelina County Lumber Company, April 20, 1935, ACLCo. Papers, Forest History Collections, Stephen F. Austin State University Library, Nacogdoches, Texas; hereafter cited as SFA Library.

[6] Fickle, *The New South and the "New Competition,"* pp. 134–39; John M. Collier, *The First Fifty Years of the Southern Pine Association, 1915–1965,* pp. 101–104; see also William G. Robbins, "The Great Experiment in Industrial Self-Government: The Lumber Industry and the National Recovery Administration," *Journal of Forest History* 25, no. 3 (July, 1981): 128–43.

of enforcement authority. Gifford Pinchot, Henry S. Graves, and Ferdinand Silcox, as chiefs of the United States Forest Service, had advocated cutting and restocking regulations that would be enforceable on private and public lands alike. As mentioned earlier, W. Goodrich Jones, the "Father of Texas Forestry," had advocated the rule that he had observed in practice in his study of European forests: cutting down a tree requires planting one in its place. The Southern Pine Association endorsed the conservation provisions of Article X and urged that the lumbermen of the southern-pine industry voluntarily accept these principles as their own policies. A few Texas lumbermen, such as Kurth, Temple, and J. Lewis Thompson, had pursued selective-cutting, sustained-yield, and reforestation policies on their own lands for years. Now increasing numbers of forest owners were converted to Jones's doctrine that "good forestry was good business." The legacy of the NRA lumber code in Texas was the general acceptance by the great majority of lumbermen of the merit of Article X: conservation for wise use, scientific land management, and a perpetual forest.[7]

The New Deal conservation programs directly affected Texas and Texas forests in many ways. One of the first emergency acts passed by Congress in the spring of 1933 gave the president broad powers to recruit unemployed young men and put them to work at conservation tasks on the public lands. By executive order Roosevelt immediately created the Civilian Conservation Corps (CCC). Within a month the program was functioning, and CCC camps sprang up in every state and territory. The inspiration for the CCC has been claimed for many groups. New York, Washington, and California and several European countries had established forest work camps for unemployed

young men before 1933. The CCC was, however, essentially FDR's own idea, one that he had been promoting, off and on, for many years. As presidential candidate and president-elect he had traveled over the country and noted the millions of acres of abandoned farms, neglected fields, cutover land grown up in worthless brush, and waste. It was clear that conservation and unemployment relief would march hand in hand in combating the Great Depression. In establishing the Civilian Conservation Corps, FDR took the young men of the country and the land, both largely wasted resources, and tried to save both.[8]

Most of the CCC's forestry work was expected to be on public lands, both federal and state, but in 1933 the Texas government held little forested land, and the federal government held none in the state. Thanks to the efforts of the Texas Forest Service under E. O. Siecke, the state had acquired about 7,000 acres in forested land in East Texas (the tracts later to be designated as state forests honoring Jones, Siecke, Fairchild, and Kirby). This acreage had been used largely for forestry-management demonstration plots and for pine-seedling nurseries.[9] Although the Texas State Forests could furnish only a small base for a conservation program, additional projects such as fire prevention and soil-erosion control on private lands, flood control, and tree planting along the state's highways provided work for the CCC "army." Under the direction of the Texas Forest Service four CCC camps were established in the spring of 1933, and by the end of the year the number had grown to eighteen. During the next two years about 4,000 CCC workers laid out more than 2,000 miles of forest roads and fire lanes, constructed 838 miles of telephone lines, erected 44 fire towers, and spent almost 21,500 man-days fighting forest

[7] Greeley, Forests and Men, pp. 137–38; Fickle, The New South and the "New Competition," pp. 260–62; Robert S. Maxwell, "One Man's Legacy: W. Goodrich Jones and Texas Conservation," Southwestern Historical Quarterly 77 (January, 1974):356–80.

[8] John A. Salmond, The Civilian Conservation Corps, 1933–1942: A New Deal Case Study, pp. 4–5.

[9] Robert S. Maxwell and James W. Martin, A Short History of Forest Conservation in Texas, 1880–1940, Stephen F. Austin State University, School of Forestry Bulletin no. 20 (1970), pp. 33–35.

A new second-growth forest. *Courtesy of U.S. Forest Service.*

fires. As could be anticipated, losses to forest fires decreased dramatically.[10]

Because Texas had retained its public lands when it became a state in 1845, the federal government held neither national parks nor national forests in the Lone Star State at the time of the Great Depression. National forestry leaders, including the chief of the Forest Service, had urged that the United States acquire lands in the states east of the Rocky Mountains, especially in cutover areas, as authorized by the Clarke-McNary Act of 1924 and recommended in the Copeland report, published in 1932. On such lands, which

[10] Ibid., pp. 35–37; Texas Forest Service, *Eighteenth and Nineteenth Annual Reports of the State Forester*; Texas A&M College, Texas Forest Service Bulletin no. 25 (1935).

would become national forests, the government would promote conservation practices and encourage multiple use by the public. Thus in 1933 the federal government was ready to purchase cutover lands in states where no substantial national forests existed, and Congress was prepared to provide the funds for their acquisition.

In Texas large acreages of cutover land were available for purchase, and there was a general willingness for the federal government to establish national forests in East Texas. Many companies had cut out and discontinued operations, and their lands, once the chief base of county-government support, had become delinquent on the tax rolls, with no purchasers in view. Residents of these counties learned that under national forest regulations the Forest

A young second-growth forest. *Courtesy of U.S. Forest Service.*

Service would return 25 percent of the sale price of standing timber to the local government in lieu of taxes. Both the lumber companies and the local government officials in East Texas were eager for the creation of national forests to promote conservation and recovery.

To this end the Texas legislature acted promptly in May, 1933, passing a bill sponsored by State Senator John S. Redditt of Lufkin and supported by conservation groups and influential lumbermen throughout the state. The measure authorized the federal governments to purchase lands in Texas for conservation purposes and ordered the state forester, Siecke, to determine the purchase-area boundaries of four national forests. Representatives of the United States Forest Service appraised the lands offered for sale within the designated boundaries and purchased those offered at acceptable prices. The surveying, appraisal, and purchase process was carried out in 1935 and early 1936. The federal government bought more than 90 percent of the lands that were to form the Texas national forests from eleven lumber companies. They were happy to find a purchaser for their cutover lands. The principal companies involved and the average prices paid were as follows:

Angelina National Forest

Kirby Lumber Company	57,035 acres
Long-Bell Lumber Company	73,880
Pickering Lumber Company	3,922
Cameron Lumber Company	14,116
Total	148,943 acres

(average price: $2.91 per acre)

Davey Crockett National Forest

Houston County Timber Company	94,126 acres
Trinity County Lumber Company	61,419
Total	155,545 acres

(average price: $8.90 per acre)

Sabine National Forest

Pickering Lumber Company	85,699 acres

Temple Lumber Company	80,974
Gilmer Lumber Company	12,509
Total	179,182 acres

(average price: $2.82 per acre)

Sam Houston National
Forest

Delta Land and Timber Company	82,774 acres
Foster Lumber Company	32,183
Gibbs Brothers and Company	30,440
Total	145,397 acres

(average price: $4.00 per acre)

The United States Forest Service acquired additional acreage from private owners bringing the total to about 660,000 acres within an area of 1,716,964 acres. In October, 1936, President Roosevelt officially proclaimed the Texas national forests and gave them their names, which "reflect the history and geography of Texas." Soon the United States Forest Service began administering the Texas national forests, under the first supervisor, Loren L. Bishop, with the avowed purposes of (1) protecting the sustained-yield concept of forest products for the benefit of local residents and industries, (2) developing recreational facilities for public use and enjoyment, (3) experimenting with and demonstrating the latest approved forestry practices, and (4) protecting watersheds for the prevention of erosion.[11]

With the establishment of the Texas national forests the United States Forest Service also took over the direction of several of the CCC camps and began the task of cleaning up and opening the newly acquired acreage. A major project was the planting of pine seedlings. Indeed, it was said that "Roosevelt's Tree Army" planted more trees than had all earlier groups put together. That was true in Texas as well as elsewhere in the United States. In the Angelina

National Forest, 23,500 acres were planted to pine during these years; 72 million trees were planted in Texas. The Texas Forest Service continued to direct CCC camps not connected with the national forests. Some camps were moved, and others were discontinued, but the Texas Forest Service was still operating five CCC camps as late as 1940. The conservation work accomplished by these young men was impressive. For example, by 1940 they had erected 81 fire towers, built almost 3,000 miles of forest roads, laid more than 2,000 miles of telephone lines, and contributed more than 115,000 man-days to fighting forest fires. Truly the work of the Civilian Conservation Corps and the development of the Texas national forests played a large part in the revival of the Texas forest industry by the end of the 1930s.[12]

Perhaps the most significant development relating to the Texas forests in the 1930s was the success of the "Herty gamble"—the production of satisfactory newsprint from southern yellow pine. This accomplishment revolutionized both the newspaper and the forest-products industries in the South and helped pull many lumbermen out of the Depression. As early as 1910 southern mills were producing kraft paper, primarily for wrapping paper and paper bags. Better grades of paper, however, especially newsprint, seemed to be beyond the capacity of southern mills because of the high resin content of pine timber. The center of the paper industry in 1930 was in the Northeast and Canada, and the Canadian share of the market was steadily increasing. By 1930 imports of paper stocks, insignificant in earlier years, amounted to two-thirds of the newsprint tonnage consumed by American pa-

[11] Maxwell and Martin, *A Short History of Forest Conservation in Texas*, pp. 35–39; John Courtenay, "Texas National Forests Are 40 Years Old," *Texas Forestry*, April, 1974; *Texas Almanac, 1958–1959*, pp. 195–96.

[12] Texas Forest Service, *Twenty-second and Twenty-third Annual Reports, 1937–1940*, Texas A&M College, Texas Forest Service Bulletin no. 27 (1938); Texas Forest Service, *Twenty-fourth and Twenty-fifth Annual Reports, 1939–1940*, Texas A&M College, Texas Forest Service Bulletin no. 30 (1940); Nixon, ed., *Franklin D. Roosevelt and Conservation*, 1:591–93; Salmond, *The Civilian Conservation Corps*, pp. 121–25.

Charles H. Herty. *Courtesy of Stephen F. Austin State University, Forest History Collections.*

Ernest Kurth. *Courtesy of Stephen F. Austin State University, Forest History Collections.*

pers. If the United States had any hope of freeing itself from Canadian dominance of paper sales, American manufacturers must find a cheap source of suitable wood.[13]

Wood scientists had worked on this problem since the turn of the century. The United States Forest Products Laboratory in Madison, Wisconsin, had conducted experiments in the 1920s that showed promise, but the paper produced retained a dark color. The scientists concluded that even if a method was developed to bleach the paper the laboratory process was too expensive for commercial use.[14]

About this time Charles H. Herty turned his attention and energies to the making of newsprint from southern yellow pine. Herty, an established chemist with a doctorate from Johns Hopkins University, had served as president

of the American Chemical Society and editor of the *Journal of Industrial and Engineering Chemistry.* In 1930 he returned south to establish a laboratory in Savannah, Georgia. With financial assistance from both public and private sources, he developed a process that produced a satisfactory newsprint pulp in his laboratory. He tested its commercial possibilities by sending about twenty-five tons of the pulp to a paper mill in Ontario and persuaded nine Georgia newspapers to use the resulting newsprint in special editions. The results were highly successful, and the newspaper owners were enthusiastic about the product. Pressmen commented that the newsprint went through the machines without a single break, and they were delighted with the print.[15]

Despite his laboratory success, Herty had

[13] David C. Smith, *History of Papermaking in the United States, 1691–1969,* pp. 219–87, 391–410.

[14] Charles A. Nelson, *History of the U.S. Forest Products Laboratory,* pp. 117–20.

[15] Jack P. Oden, "Charles Holmes Herty and the Birth of the Southern Newsprint Paper Industry, 1927–1940," *Journal of Forest History* 21, no. 2 (April, 1977): 77–81; *Time,* March 27, 1933, p. 24.

Southland Paper Mills under construction. *Courtesy of Stephen F. Austin State University, Forest* *History Collections.*

great difficulty attracting financial backing to put his process into industrial production. Many northern and Canadian paper-mill executives resented Herty's efforts to establish a newsprint industry in the South. They argued that the industry was already overcrowded and that development of southern paper mills could be accomplished only at the expense of the established mills in the Northeast. During the brief period of the NRA, spokesmen for the American Paper and Pulp Association effectively blocked all efforts to establish a paper mill in the South and forestalled the efforts of Herty and his friends to obtain loans from the Reconstruction Finance Corporation, another federal Depression-fighting agency. Herty, however, proved to be an active and able publicist. He traveled, spoke, and wrote pamphlets extolling the advantages of a southern newsprint industry to both timber producers and paper users. He sought to clear up inaccurate and exaggerated press stories about his experiments. He pointed out the cost differentials, showing that newsprint produced from southern pine would be one-third cheaper than that produced in the North. In spite of these efforts and continued laboratory experiments that

further improved the paper pulp, from 1933 to 1936 Herty had no success in establishing a commercial newsprint industry in the South.[16]

Then Herty met Ernest L. Kurth at an industrial conference. The scientist explained his experiments in making newsprint and the possibilities it provided for transforming the timber industry of the South. Herty complained that he could not find anyone "bold" enough to take the risk to develop the process commercially. At this time Kurth headed a large, varied industrial complex in East Texas centering in the Angelina County Lumber Company. He prided himself as an entrepreneur in the classical sense (a risk taker) and a decision maker. Herty's story challenged Kurth, and after visits to Herty's laboratory in Savannah and conversations with other industrialists and publishers, he decided to build a mill in Texas to manufacture newsprint. He found that friends and acquaintances were leery of investing in such a speculative enterprise. They pictured Herty as a visionary and told Kurth that he was

[16] Oden, "Charles Holmes Herty and the Birth of the Southern Newsprint Paper Industry, 1927–1940," pp. 82–85: *Lufkin* (Texas) *News*, October 21, 1962.

Paper machine, Southland Paper Mills. A dream realized, a gamble won. *Courtesy of Stephen F. Aus-*
tin State University, Forest History Collections.

a fool—"and would soon be a bankrupt fool."
Nevertheless, Kurth persisted. He interested
James G. Stahlman, of the *Nashville Banner,*
and E. M. Dealey, of the *Dallas Morning*
News, in the project, and they promoted it
with other southern and southwestern publishers. By the summer of 1937 they were able
to tell Kurth that the publishers would take
all the newsprint he could make during the
first five years. Louis F. Calder, of Perkins-
Goodwin Company, a New York paper company, pledged financial support for the venture.
Kurth also got financial backing and participation from other Texas lumbermen, including
Arthur Temple. With these promises of firm
support, Kurth then went to Washington, D.C.,
and persuaded fellow Texan Jesse Jones, chairman of the Reconstruction Finance Corporation, to advance an additional loan to the
enterprise.[17]

The Southland Paper Mills, Inc., as the com-

pany was called, came into being in June,
1938, with Kurth as president, Arthur Temple
as vice-president, and Simon Henderson, Jr., as
secretary. They began construction of a mill
at Lufkin early in 1939, and on January 17,
1940, the Southland Paper Mills began the
first production of newsprint from southern
yellow pine. It grew into a multimillion-dollar
enterprise, and additional units were added to
the mill at Lufkin. Other paper mills sprang up
in Texas and other parts of the South. Eventually, almost as much pine timber went into
pulp for the paper mills as into saw lumber.
The new process gave the southern forest owners and tree farmers a new, profitable outlet for
their young pine trees.[18]

The "Herty gamble" paid off manyfold.
Herty was honored as a scientific pioneer, and
Kurth emerged as an industrial statesman with

[17] *Lufkin News,* October 21, 1962; Oden, "Charles Holmes
Herty and the Birth of the Southern Newsprint Paper Industry,
1927–1940," pp. 86–89.

[18] *Texas Forestry* 14, no. 4 (April, 1974). There is considerable correspondence relating to the paper-mill project in the papers of the Angelina County Lumber Company between 1936
and 1940. See, for example, Arthur Temple to Kurth, June 15,
1938; Arthur Temple to Kurth, August 11, 1938; Albert New-
combe (Perkins-Goodwin Company) to Kurth, January 24,
1939. ACLCo. Papers, Forest History Collections, SFA Library.

Arthur Temple, Jr., the third-generation heir of Temple lumber enterprises, was a young man in 1940. *Authors' collections.*

broad vision whose proposals few henceforth would challenge. It was, indeed, a major milestone in the history of Texas forests and forest-products industries.

Texas and Texas lumbermen benefited from yet another New Deal program. At Roosevelt's urging Congress created the Federal Housing Authority, which constructed housing units and made loans at low interest rates to individuals who wished to build homes. The measure stimulated the construction industry, which had been prostrated by the Great Depression. In turn the Texas lumber manufacturers began to revive, and production slowly increased to pre-Depression levels. Workers' benefits increased also. The eight-hour day, the federal minimum wage, and cash wages (replacing the hated merchandise checks) became the rule in the Texas lumber industry, especially in the larger, permanent mills. By 1940 lumber production in Texas had once more passed 1 billion board feet a year, about equal to the production of the last pre-Depression year, 1929. The Texas lumber industry had emerged from the depths of the Depression with a strong base and a commitment to conservation and good forestry practices. Both large and small millowners looked to the future with optimism.[19]

What would have been the history of the Texas forests and the forests workers' after 1940 under peacetime conditions would be pure speculation. By 1940, however, World War II was raging in Europe and the Far East. Within a year the United States had entered the conflict, and all efforts were bent to achieve victory over the Axis Powers. Millworkers, loggers, and the young men of the CCC who had planted trees and fought forest fires were soon in the armed forces. Conservation, sustained-yield, and reforestation practices, home construction, and application of new technologies would have to wait until the struggle had ended.

[19] *Texas Almanac, 1958–1959*, p. 190.

A vigorous second-growth forest in postwar Texas.
Authors' collections.

Bibliography

Primary Sources

Angelina County Lumber Company (ACLCo.) Papers, 1887–1966. Includes papers of subsidiary companies and the Angelina and Neches River Railroad Papers. Forest History Collections, Stephen F. Austin State University Library, Nacogdoches.

Blake, R. B. Research Collection (typescript), Nacogdoches Archives. 75 vols. plus 18 supplements. Compiled in Eugene C. Barker Texas History Center, Austin, 1958–59. Copies in Stephen F. Austin State University Library, Nacogdoches.

Carter, W. T., and Brother Lumber Company Records. Camden.

Forest History Collections, Stephen F. Austin State University Library, Nacogdoches. Miscellaneous Papers: Oral History Collections, interviews, reminiscences, letters, maps, and other items.

Foster Lumber Company Papers. Forest History Collections, Stephen F. Austin University Library, Nacogdoches.

Frost-Johnson Lumber Company Papers. Includes papers of the Nacogdoches and Southeastern Railroad. Forest History Collections, Stephen F. Austin State University Library, Nacogdoches.

Gilmer, Alexander, Lumber Company Papers. University of Texas Archives, Austin.

Jones, W. Goodrich, Papers. Forest History Collections, Stephen F. Austin State University Library, Nacogdoches.

Kirby, John Henry, Personal Papers. University of Houston Library, Houston.

Kirby Lumber Corporation Papers. Forest History Collections, Stephen F. Austin State University Library, Nacogdoches.

Kurth Family Papers. In possession of Dr. Robert L. Kurth, Lufkin, Texas.

Lutcher and Moore Lumber Company Papers. Forest History Collections, Stephen F. Austin State University Library, Nacogdoches.

Ragan, Cooper K., Papers. University of Houston Library, Houston.

Speek, Peter A. "Notes on Investigations of Three Texas Lumber Towns." United States Department of Labor, Reports of the Commission on Industrial Relations, 1914. Microfilm. Stephen F. Austin State University Library, Nacogdoches.

Temple Industries Papers. Forest History Collections, Stephen F. Austin State University Library, Nacogdoches.

Federal Government Publications

Blanchard, Newton C., et al., eds. *Proceedings of a Conference of Governors in the White House, Washington, D.C., May 13–15, 1908.* Washington, D.C.: Government Printing Office, 1909.

Bray, William L. *Forest Resources of Texas.* U.S. Department of Agriculture, Bureau of Forestry Bulletin no. 47. Washington, D.C.: Government Printing Office, 1904.

Department of Agriculture. *Trees: The Yearbook of Agriculture, 1949.* Washington, D.C.: Government Printing Office, 1949.

Department of Commerce, Bureau of the Census. *United States Census, 1900.* Washington, D.C.: Government Printing Office, 1901.

———. *United States Census, 1910, Supplement for Texas.* Washington, D.C.: Government Printing Office, 1911.

———. *United States Census, 1920.* Washington, D.C.: Government Printing Office, 1921.

———. *United States Census, 1940, Supplement for Texas.* Washington, D.C.: Government Printing Office, 1941.

Department of Labor, Bureau of Labor Statistics. *Industrial Survey in Selected Industries in the United States, 1919.* Bureau of Labor Statistics Bulletin no. 265. Washington, D.C.: Government Printing Office, 1920.

———. *Wages and Hours of Labor, 1916; 1929.* Washington, D.C.; Government Printing Office, 1916; 1930.

Interstate Commerce Commission, Bureau of Statistics, *Interstate Commerce Commission Activities, 1887–1937.* Washington, D.C.: Government Printing Office, 1937.

Mohr, Charles. *The Timber Pines of the Southern United States.* U.S. Department of Agriculture, Division of Forestry Bulletin no. 18. Washington, D.C.: Government Printing Office, 1897.

Nelson, Charles A. *History of the U.S. Forest Products Laboratory.* Washington, D.C.: U.S. Department of Agriculture, 1971.

Nixon, Edgar B., ed. *Franklin D. Roosevelt and Conservation.* 2 vols. Washington, D.C.: Government Printing Office, 1957.

Steer, Henry B. *Lumber Production in the United States, 1799–1946.* Washington, D.C.: Government Printing Office, 1948.

U.S. Shipping Board. *Fourth Annual Report of the United States Shipping Board.* Washington, D.C.: Government Printing Office, 1920.

United States Reports

Houston, East and West Texas Railway v. *United States.* 234 U.S. 342 (1914).

O'Keef v. *United States.* 240 U.S. 294 (1915).

United States v. *Louisiana and Pacific Railroad Company et al.* 234 U.S. 1, 1185–1198 (1913).

State of Texas Publications

Clark, Charles E. "The Texas Workmen's Compensation Act." *Vernon's Annotated Statutes.* Vol. 22. Kansas City, Mo.: Vernon Law Book Company, 1958.

Gammel, H. P. N. *Laws of Texas, Supplement.* 14 vols. Austin, 1899–1937.

House Journal of the Texas Legislature. 34th Leg., reg. sess., 1915. Austin, 1915.

Senate Journal of the Texas Legislature. 34th Leg., reg. sess., 1915. Austin, 1915.

Texas Bureau of Labor Statistics. *Biennial Report, 1927–1928.* Austin, 1928.

Texas Forest Service Publications

Foster, J. H. *Second Annual Report of the State Forester.* Texas A&M College, Department of Forestry Bulletin no. 8, 1917.

———, Krausz, H. B., and Johnson, George W. *First Annual Report of the State Forester.* Texas A&M College, Department of Forestry Bulletin no. 4, 1917.

———, ———, and ———. *Forest Resources of Eastern Texas.* Texas A&M College, Department of Forestry Bulletin no. 5, 1917.

———, ———, and Leidigh, A. H. *General Survey of Texas Woodlands, including a Study of the Commercial Possibilities of Mesquite.* Texas A&M College, Department of Forestry Bulletin no. 3, 1917.

Texas Forest Service. *East Texas Protection Area Forest Fire Statistics.* Texas A&M College, Texas Forest Service Bulletin, 1957.

———. *Eighteenth and Nineteenth Annual Reports of the State Forester.* Texas A&M College, Texas Forest Service Bulletin no. 25, 1935.

———. *Twenty-second and Twenty-third Annual Reports, 1937–1938.* Texas A&M College, Texas Forest Service Bulletin no. 27, 1938.

———. *Twenty-fourth and Twenty-fifth Annual Reports, 1939–1940.* Texas A&M College, Texas Forest Service Bulletin no. 30, 1940.

Texas Superior Court Report

Ed Jordon v. *Texas.* 51 *Tex. Crim. Rpt.* 531 (1907).

Books and Pamphlets

Abernethy, Francis E. *Tales from the Big Thicket.* Austin: University of Texas Press, 1966.

Acheson, Sam Hanna. *Joe Bailey: The Last Democrat.* New York: Macmillan Co., 1932.

Adams, Karmer A. *Logging Railroads of the West.* Seattle, Wash.: Superior Publishing Co., 1961.

Allen, Ruth A. *East Texas Lumber Workers: An Economic and Social Picture, 1870–1950.* Austin: University of Texas Press, 1961.

Andrews, Ralph W. *This Was Sawmilling.* Seattle, Wash.: Superior Publishing Co., 1957.

Beebe, Lucius. *Mixed Train Daily*. New York: E. P. Dutton and Co., 1947.

Bollaert, William. *William Bollaert's Texas*. Edited by W. Eugene Hollon and Ruth Lapham Butler. Norman: University of Oklahoma Press, 1956.

Botkin, B. A., and Harlow, Alvin F., eds. *A Treasury of Railroad Folklore: The Stories, Tall Tales, Traditions, Ballads, and Songs of the American Railroad Man*. New York: Crown Publishers, 1953.

Bowman, Bob. *This Was East Texas: An Anthology of Ghost Towns*. Diboll, Tex.: Angelina Free Press, 1966.

Brown, Arthur A., and Folweiler, A. D. *Fire in the Forests of the United States*. St. Louis, 1946. Planographed by John S. Swift Co., 1953.

Brown, John Henry. *Indian Wars and Pioneers of Texas*. Austin: L. E. Daniell, n.d.

Brown, Nelson Courtlandt. *Logging: Principles and Practices in the United States and Canada*. New York: John Wiley and Sons, 1934.

———. *Lumber: Manufacture, Conditioning, Grading, Distribution, and Use*. New York: John Wiley and Sons, 1947.

Bryant, Claude W. *Lumbering Along in Texas*. San Antonio, Tex.: Naylor Co., 1960.

Bryant, Keith L., Jr. *History of the Atchison, Topeka, and Santa Fe Railway*. New York, Macmillan Co., 1974.

Bryant, Ralph C. *Logging: The Principles and General Methods of Operation in the United States*. New York: John Wiley and Sons, 1914.

———. *Lumber: Its Manufacturing and Distribution*. New York: John Wiley and Sons, 1922.

Carroll, Benajah Harvey. *Standard History of Houston, Texas, from a Study of the Original Sources*. Knoxville, Tenn.: H. W. Crew, 1912.

Clark, James A., and Halbouty, Michael T. *Spindletop*. New York: Random House, 1952.

Cochran, Thomas C., and Miller, William. *The Age of Enterprise: A Social History of Industrial America*. 2d ed. New York: Harper & Brothers, 1961.

Collier, John M. *The First Fifty Years of the Southern Pine Association, 1915–1965*. New Orleans: Southern Pine Association, 1965.

Commons, John R.; Brandeis, Elizabeth; Perlman, Selig; and Taft, Philip. *History of Labor in the United States, 1896–1932*. 4 vols. New York: Macmillan Co., 1935.

Crocket, George L. *Two Centuries in East Texas: A History of San Augustine County and Surrounding Territory from 1685 to the Present Time*. Dallas: Southwest Press, 1932.

Daugherty, Carroll R. *Labor Problems in American Industry*. New York: Houghton Mifflin Co., 1948.

Deal, Winnie Mims. *Jefferson, Texas: Queen of the Cypress*. Dallas: Mathis, Van Nort Co., 1953.

Defebaugh, James E. *American Lumbermen*. 3 vols. Chicago: American Lumberman, 1908.

Dionne, Jack. *A Brief Story of the Life of John Henry Kirby*. Houston: Privately printed by the employees of Kirby Lumber Co., n.d.

Dobie, J. Frank. *The Flavor of Texas*. Dallas: Dealey and Lowe, 1936.

Faulkner, Harold U. *Politics, Reform, and Expansion, 1890–1900*. New York: Harper Brothers, 1959.

Fickle, James E. *The New South and the "New Competition": Trade Association Development in the Southern Pine Industry*. Urbana: University of Illinois Press, 1980.

Fries, Robert F. *Empire in Pine: The Story of Lumbering in Wisconsin, 1830–1900*. Madison: State Historical Society of Wisconsin, 1951.

Greeley, William B. *Forests and Men*. New York: Doubleday and Co., 1951.

Grodinsky, Julius. *Jay Gould: His Business Career, 1867–1892*. Philadelphia: University of Pennsylvania Press, 1957.

Hickman, Nollie. *Mississippi Harvest: Lumbering in the Longleaf Pine Belt, 1840–1915*. University: University of Mississippi Press, 1962.

Hidy, Ralph W.; Hill, Frank E.; and Nevins, Allan. *Timber and Men: The Weyerhaeuser Story*. New York: Macmillan Co., 1963.

Holbrook, Stewart H. *Burning an Empire*. New York: Macmillan Co., 1960.

———. *Holy Old Mackinaw*. New York: Macmillan Co., 1938.

Holley, Mary Austin. *Texas*. Lexington, Ky., 1836. Reprint. Austin: Steck Co., 1935.

Horn, Stanley F. *This Fascinating Lumber Business*. New York: Bobbs-Merrill Co., 1959.

Hurley, Edward N. *The Bridge to France*. Philadelphia: J. B. Lippincott Co., 1927.

———. *The New Merchant Marine*. New York: Century Co., 1920.

Jensen, Vernon H. *Lumber and Labor.* New York: Farrar and Rinehart, 1945.

Kerr, Ed. *History of Forestry in Louisiana.* Baton Rouge, La.: Office of the State Forester, 1958.

King, John O. *Early History of the Houston Oil Company of Texas, 1901–1908.* Houston: Texas Gulf Coast Historical Association, 1959.

Labbe, John T., and Goe, Vernon. *Railroads in the Woods.* Berkeley: University of California Press, 1961.

Lasswell, Mary. *John Henry Kirby: Prince of the Pines.* Austin: Encino Press, 1967.

Lathrop, Barnes F. *Migration into East Texas, 1835–1860.* Austin: Texas State Historical Association, 1949.

Lay, Bennett. *The Lives of Ellis P. Bean.* Austin: University of Texas Press, 1960.

Leclerc, Frédéric. *Texas and Its Revolution.* Translated and edited by James L. Shepherd III. Houston: A. Jones Press, 1950.

Lillard, Richard G. *The Great Forest.* New York: Alfred A. Knopf Co., 1947.

Link, Arthur S. *American Epoch.* New York: Alfred A. Knopf Co. 1955.

McClelland, John M., Jr. *R. A. Long's Planned City: The Story of Longview.* Longview, Wash.: Longview Publishing Co., 1976.

McCleskey, Clifton; Dickens, E. Larry; and Butcher, Allan K. *The Government and Politics of Texas.* Boston: Little, Brown and Co., 1966.

McKay, Mrs. Arch, and Spellings, Mrs. H. A. *A History of Jefferson.* 5th ed. Jefferson, Tex., 1964.

McReynolds, Edwin C. *Missouri: A History of the Crossroads State.* Norman: University of Oklahoma Press, 1962.

Maissin, Eugène. *The French in Mexico and Texas, 1838–1839.* Translated with an introduction by James L. Shepherd III. Salado, Tex.: A. Jones Press, 1961.

Masterson, Vincent V. *The Katy Railroad and the Last Frontier.* Norman: University of Oklahoma Press, 1952.

Mattox, W. C. *Building the Emergency Fleet.* Cleveland, Ohio: Penton Publishing Co., 1920.

Maxwell, Robert S. *La Follette and the Rise of the Progressives in Wisconsin.* Madison: State Historical Society of Wisconsin, 1956.

———. *Whistle in the Piney Woods: Paul Bremond and the Houston, East and West Texas Railway.* Houston: Texas Gulf Coast Historical Association, 1963.

———, and Martin, James W. *A Short History of Forest Conservation in Texas, 1880–1940.* Stephen F. Austin State University, School of Forestry Bulletin no. 20, 1970.

Muir, Andrew Forest, ed. *Texas in 1837: An Anonymous Contemporary Narrative.* Austin: University of Texas Press, 1958.

Nordyke, Lewis. *The Truth about Texas.* New York: Crowell Publishers, 1957.

Olmsted, Frederick Law. *The Cotton Kingdom.* Reprint. New York: Alfred A. Knopf Co., 1953.

———. *A Journey through Texas.* New York, 1857. Reprint. Edited by James Howard. Austin: von Boeckman–Jones Press, 1962.

Pinchot, Gifford. *Breaking New Ground.* New York: Harcourt, Brace, Publishers, 1947.

Reed, St. Clair G. *A History of the Texas Railroads and of the Transportation Conditions under Spain and Mexico and the Republic and State.* Houston: St. Clair Publishing Co., 1941.

Richardson, Rupert N.; Wallace, Ernest; and Anderson, Adrian. *Texas: The Lone Star State.* 4th ed. Englewood Cliffs, N.J.: Prentice-Hall, 1981.

Richardson, Thomas C. *East Texas: Its History and Its Makers.* 4 vols. New York: Lewis Historical Publishing Co., 1940.

Roach, Hattie Joplin. *A History of Cherokee County, Texas.* Dallas: Southwest Press, 1937.

Salmond, John A. *The Civilian Conservation Corps, 1933–1942: A New Deal Case Study.* Durham, N.C.: Duke University Press, 1967.

Schlesinger, Arthur M., Jr. *The Coming of the New Deal.* Boston: Houghton Mifflin Co., 1958.

Smith, David C. *History of Papermaking in the United States, 1691–1969.* New York: Lockwood Publishing Co., 1970.

Southern Pine Association. *Lumber Production in the South.* New Orleans, 1963.

Southern Pine Inspection Bureau. *Short Course in Grading.* New Orleans: Southern Pine Inspection Bureau, 1963.

Sperry, Marcus E. *The Port of Orange, Texas, U.S.A.* Orange, 1916.

Spratt, John S. *The Road to Spindletop: Economic Change in Texas, 1875–1901.* Dallas: Southern Methodist University Press, 1955.

Stratton, Florence. *Story of Beaumont.* Houston: Hercules Printing and Book Co., 1925.

Taber, Thomas T. III, and Casler, Walter. *Climax: An Unusual Steam Locomotive.* Rahway, N.J.: Railroadians of America, 1960.

Texas State Federation of Labor. *Proceedings, Seventh Annual Convention, 1904.* Galveston, 1904.

Tolson, R. J. *A History of William Cameron and Co., Inc.* Waco, Texas: Privately printed, 1920.

Wahlenberg, William G. *Longleaf Pine: Its Use, Ecology, Regeneration, Protection, Growth, and Management.* Washington, D.C.: Charles Lathrop Pack Forestry Foundation, 1946.

Watson, Mrs. James. *The Lower Rio Grande Valley of Texas and its Builders.* Mission, Tex: Privately published, 1931.

Webb, Walter P., editor-in-chief. *The Handbook of Texas.* 3 vols. Austin: Texas State Historical Association, 1952–76.

Wertenbaker, Green Payton. *The Face of Texas: A Survey in Words and Pictures.* New York: Crowell Publishers, 1961.

Who's Who in the South and Southwest, 1973–1974. 13th ed. Chicago: Marquis Who's Who, 1973.

Wilson, Thomas A. *Some Early Southeast Texas Families.* Edited by Madeline Martin. Houston: Lone Star Press, 1965.

Winkler, Ernest W. *Platforms of Political Parties in Texas.* University of Texas Bulletin no. 53, 1916.

Wright, Solomon Alexander. *My Rambles as an East Texas Cowboy, Hunter, Fisherman, Tie-Cutter.* Austin: Texas Folklore Society, 1942.

Zies, Paul M. *American Shipping Policy.* Princeton, N.J.: Princeton University Press, 1938.

Zlatkovich, Charles P. *Texas Railroads: A Record of Construction and Abandonment.* Austin: University of Texas, Bureau of Business Research, 1981.

Articles

Appel, John C. "Regionalism and American History." *Social Education* 13 (November, 1949): 319–24.

Barker, Eugene Campbell, ed. "Stephen F. Austin's Descriptions of Texas." *Southwestern Histori-cal Quarterly* 28 (October, 1924): 98–104.

Bowman, Bob. "Touring East Texas." *Daily Sentinel* (Nacogdoches), September 16, 1965.

Boyd, James. "Fifty Years in the Southern Pine Industry." *Southern Lumberman* 144 (December 15, 1931): 59–67.

Buckley, S. B. "Pine Lands of Southeastern Texas." In *Texas Almanac, 1868.* Galveston, 1868.

Chambers, William T. "Divisions of the Pine Forest Belt of East Texas." *Economic Geography* 6 (January, 1930): 94–103.

———. "Geographic Regions of Texas." *Texas Geographic Magazine,* Spring, 1948, pp. 7–15.

———. "Life in a Southern Sawmill Community." *Journal of Geography* 30 (May, 1931): 181.

———. "Pine Woods Region in Southeastern Texas." *Economic Geography* 10 (July, 1934): 302–18.

———. "The Redlands of Central Eastern Texas." *Texas Geographic Magazine,* Autumn, 1941, pp. 1–15.

Clepper, Henry. "Crusade for Conservation." *American Forests* 81, no. 10 (October, 1975).

Conlin, Joseph R. "Old Boy, Did You Get Enough of Pie?" *Journal of Forest History* 23 (October, 1979): 164–85.

Courtenay, John. "Texas National Forests are 40 Years Old." *Texas Forestry,* April, 1974, pp. 18–19.

Creel, George. "The Feudal Towns of Texas." *Harper's Weekly* 60 (January 23, 1915): 76–78.

Crouch, W. T. "The Negro in the South." In *Culture in the South.* Chapel Hill: University of North Carolina Press, 1935.

Doree, Bill. "Texas' First Steam-powered Sawmill." *Gulf Coast Lumberman* 100 (April, 1963): 13.

Dugas, Vera Lea. "Texas Industry, 1860–1880." *Southwestern Historical Quarterly* 59, no. 2 (October, 1955): 161.

Fescue, Edwin J. "East Texas: A Timbered Empire." *Journal of the Graduate Research Center, Southern Methodist University* 28 (April, 1960): 1–57.

Gilman, F. H. "History of the Development of Sawmill and Woodworking Machinery." *Mississippi Valley Lumberman* 36 (February 1, 1895): 60.

Haislet, John A. "Texans Evolve a State Forestry Agency." *Texas Forests and Texans* 5 (May–June, 1964): 6–7.

"He Was the Plowboy of Peachtree Village but He Became Prince of the Pines." *East Texas*, December, 1926.

"The House of Thompson." *American Lumberman*, September 26, 1908, pp. 67–149.

Johnson, Arthur M. "The Early Texas Oil Industry: Pipelines, and the Birth of an Integrated Oil Industry, 1901–1911." *Journal of Southern History* 32 (November, 1966): 516–28.

Kerr, Ed. "Southerners Who Set the Woods on Fire." *Harper's* 217 (July, 1958): 28–33.

"The Kirby Story: 50th Anniversary of the Founding of a Lumber Empire." *Gulf Coast Lumberman* 39, no. 8 (July 15, 1951): 27–42.

Kuykendall, J. H. "Reminiscences of Early Texans." *Quarterly of the State Historical Society of Texas* 7 (July, 1903): 29–64.

"A Look at the Past." *American Lumberman*, January 18, 1908. Reprint. Diboll, Texas: Free Press, March, 1969.

Looscan, Adele. "Harris County, 1822–1845." *Quarterly of the State Historical Society of Texas* 18 (October, 1914): 195–207.

———. "Journal of Lewis Birdwell Harris." *Quarterly of the State Historical Society of Texas* 25 (January, 1922): 185–97.

———. "The Pioneer Harrises of Harris County, Texas." *Quarterly of the State Historical Society of Texas* 31 (April, 1928): 365–73.

Marvin, Winthrop L. "A Thousand Wooden Ships for War Trade." *American Review of Reviews* 55 (May, 1917): 519–20.

Maxwell, Robert S. "Lumbermen of the East Texas Frontier." *Journal of Forest History* 9 (April, 1965): 12–16.

———. "One Man's Legacy: W. Goodrich Jones and Texas Conservation." *Southwestern Historical Quarterly* 77 (January, 1974): 355–80.

———. "The Pines of Texas: A Study in Lumbering and Public Policy." *East Texas Historical Journal* 2 (October, 1964): 77–86.

"Mayhem in the Woods." *Gulf Coast Lumberman* 37, no. 5 (June 9, 1949): 36.

Morgan, George T., Jr. "No Compromise—No Recognition: John Henry Kirby, the Southern Pine Operators Association, and Unionism in the Piney Woods, 1906–1916." *Labor History* 10 (Spring, 1969): 193–204.

Oden, Jack P. "Charles Holmes Herty and the Birth of the Southern Newsprint Industry, 1927–1940." *Journal of Forest History* 21 (April, 1977): 76–89.

Richards, J. "A Treatise on the Construction and Operation of Woodworking Machines." *Journal of Forest History* 9, no. 4 (January, 1966): 16–23.

Robbins, William G. "The Great Experiment in Industrial Self-Government: The Lumber Industry and the National Recovery Administration." *Journal of Forest History* 25 (July, 1981): 128–43.

Seale, William. "River People." *East Texas Historical Journal* 5, no. 1 (March, 1967): 43–50.

Splawn, N. W. "A Review of Minimum Wage Theory and Practice with Special Reference to Texas." *Southwestern Political Science Quarterly* 1 (March, 1921): 339–71.

Taylor, Frank H. "Through Texas." *Harper's New Monthly Magazine* 59 (October, 1879): 703–18.

Wagoner, Edward R. "The Green Gold of Texas: Texas Forestry Association 50th Anniversary." *American Forests* 70, no. 9 (September, 1964): 26.

Dissertations and Theses

Balch, Ernest. "Biographical Sketches of Industrial Leaders in Lufkin and Angelina County." Master's research paper, Stephen F. Austin State College, Nacogdoches, 1949.

Collier, G. Loyd. "The Evolving East Texas Woodland." Ph.D. dissertation, University of Nebraska, Lincoln, 1964.

Easton, Hamilton Pratt. "A History of the Texas Lumbering Industry." Ph.D. dissertation, University of Texas, Austin, 1947.

Evans, Cleo F. "Transportation in Early Texas." Master's thesis, St. Mary's University, San Antonio, 1940.

Hansbro, Ruth. "History of San Jacinto County." Master's thesis, Sam Houston State Teachers College, Huntsville, 1940.

Lockhart, Bernice. "Navigating Texas Rivers, 1821–1900." Master's thesis, St. Mary's University, San Antonio, 1949.

McCord, Charles R. "A Brief History of the Brotherhood of Timber Workers." Master's thesis, University of Texas, Austin, 1959.

Mansell, Adele. "A History of Trinity County." Master's thesis, Sam Houston State Teachers College, Huntsville, 1941.

Martin, James W. "A History of Forest Conservation in Texas, 1900–1935." Master's thesis, Stephen F. Austin State University, Nacogdoches, 1966.

Mayberry, Lita M. "Keltys: An East Texas Sawmill Town." Master's thesis, Stephen F. Austin State College, Nacogdoches, 1948.

Mills, Robert E. "Navigation of the Trinity River." Master's thesis, Sam Houston State Teachers College, Huntsville, 1943.

Nowlin, Rankin S. "Economic Development of the Kirby Lumber Company of Houston, Texas." Master's thesis, George Peabody College, Nashville, 1930.

Scurlock, Virgie. "Ante Bellum Nacogdoches, 1846–1861." 2 vols. Master's thesis, Stephen F. Austin State College, Nacogdoches, 1954.

Weaver, Harry. "Labor Practices in the East Texas Lumber Industry." Master's thesis, Stephen F. Austin State College, Nacogdoches, 1961.

Newspapers and Magazines

American Lumberman, 1902–18.

American Lumberman's Review 48, no. 8 (April 25, 1925).

Athens (Tex.) Review, August, 1962 (film tape).

Beaumont Enterprise, December 1, 5, 16, 1917; November 10, 1940; March 20, 1949; November 6, 1955; August 21–22, 1957.

Champion (Center, Tex.), 1952.

Dallas Morning News, November 27, 1922; January 5–9, 1923; March 13, 1962.

Galveston Daily News, August 2, 7, 1877; December 14–16, 1890.

Gulf Coast Lumberman, August 15, 1914; June 1, 1915; January 1, 1917; May 15, 1917; August 1, 1918; December 1, 1924; November, 1963.

Houston Chronicle, November 13, 1901; January 11–13, 1918.

Houston Post, November 30, 1922; January 3–4, 1923; April 19–20, 1923; December 12, 1937.

Houston Telegraph, April 18, 1878; November 16, 1878.

Literary Digest, March 27, 1920.

Log of Long-Bell, March, June, July, 1919.

Lufkin Daily News, June 16–17, 1930; October 30, 1955; October 21, 1962; March 15, 1966.

Nacogdoches Daily Sentinel, January 4, 1923.

Nacogdoches News, July 30, 1877; March 16, 1882.

New York Times, April 30, 1933.

Orange Leader, October 16, 1904; May 29, 1936.

Redland Herald (Nacogdoches), November 30, 1922.

St. Louis Post-Dispatch, January 18, 1903.

San Augustine (Tex.) Tribune, December 5, 1935.

Southwest (Houston), November, 1907.

Texas Forestry, 1964–82.

Time, March 27, 1933.

Tyler Morning Telegraph, January 13, 1949.

Waco Tribune-Herald, April 26, 1964.

Almanacs

Burke's Texas Almanac and Immigrant Handbook. Houston, 1879, 1880, 1881.

Texas Almanac. Galveston, 1860, 1867, 1870.

Texas Almanac. Galveston and Dallas, 1904, 1910, 1911.

Texas Almanac and State Industrial Guide. Dallas, 1926, 1957, 1977, 1981.

Index

CPSIA information can be obtained
at www.ICGtesting.com
Printed in the USA
FSOW02n1200290117
30072FS